United States Nuclear Regulatory Commission

Protecting People and the Environment

NUREG-1860
Vol. 1

Feasibility Study for a Risk-Informed and Performance-Based Regulatory Structure for Future Plant Licensing

Main Report

Manuscript Completed: December 2007
Date Published: December 2007

M. Drouin, NRC Project Manager

Office of Nuclear Regulatory Research

ABSTRACT

The purpose of this NUREG is to establish the feasibility of developing a risk-informed and performance-based regulatory structure for the licensing of future nuclear power plants (NPPs). As such, this NUREG documents a "Framework" that provides an approach, scope and criteria that could be used to develop a set of requirements that would serve as an alternative to 10 CFR 50 for licensing future NPPs; however, this Framework is not the entire process. It is an initial phase in is to demonstrate the feasibility of such a concept, recognizing that for full implementation there will be outstanding programmatic, policy, and technical issues to be resolved. As such, this feasibility study does not represent a staff position, but rather a significant piece of research. The second phase, which involves implementation, is comprised of several, iterative steps: resolution of issues, development of draft requirements and regulations, pilots and tests, and rulemaking.

The information contained in this NUREG is intended for use by the US Nuclear Regulatory Commission (NRC) staff in developing requirements applicable to the licensing of commercial NPPs. Similar to 10 CFR 50, it covers the design, construction and operation phases of the plant lifecycle up to and including the initial stages of decommissioning (i.e., where spent fuel is still stored on-site). It covers the reactor and support systems. Fuel handling and storage are not addressed, but rather would be considered as part of implementation. The approach taken is one that integrates deterministic and probabilistic elements and builds upon recent policy decisions by the Commission related to the use of a probabilistic approach and mechanistic radioactive source terms in establishing the licensing basis.

At the highest level, the Framework has been developed from the top down with the safety expectation that future NPPs are to achieve a level of safety at least as good as that defined by the Quantitative Health Objectives in the Commission's 1986 Safety Goal Policy Statement. Criteria are then developed that utilize an integrated deterministic and probabilistic approach for defining the licensing basis and safety classification. Implementation of these criteria would require a design specific probabilistic risk assessment and would result in a design specific licensing basis. Defense-in-depth remains a fundamental part of the requirements development process and has as its purpose applying deterministic principles to account for uncertainties. Defense-in-depth has been defined as an element in NRC's safety philosophy that is used to address uncertainty by employing successive measures, including safety margins, to prevent or mitigate damage if a malfunction, accident, naturally or intentional caused event occurs. The approach taken in the Framework continues the practice of ensuring that the allowable consequences of events are matched to their frequency such that frequent events are to have very low consequences and less frequent events can have higher consequences. This is expressed in the form of a frequency-consequence curve. The allowable consequences are based upon existing dose limits, and the associated frequencies are based on guidance contained in International Commission on Radiological Protection 64 and engineering judgment.

Part of the process involves development of guidance to be used for actually writing the requirements. This guidance addresses writing the requirements in a performance-based fashion, incorporating lessons learned from past experience, and utilizing existing requirements and guidance, where practical. The guidance also ensures that the probabilistic process for establishing the licensing basis are incorporated. All of the above are integrated and results in a set of potential requirements which serve to illustrate and establish the feasibility of developing a risk-informed and performance-based licensing approach.

Paperwork Reduction Act Statement

The information collections contained in this document are subject to the Paperwork Reduction Act of 1995 (44 U.S.C. 3501 et seq.), which were approved by the Office of Management and Budget (OMB), approval numbers 3150-0011, 3150-0014, 3150-0050, 3150-0051, 3150-0093, and 3150-0197.

Public Protection Notification

The NRC may not conduct or sponsor, and a person is not required to respond to a request for information or an information collection requirement unless the requesting document displays a currently valid OMB control number.

FOREWORD

The Commission, in its Policy Statement on Regulation of Advanced Nuclear Power Plants, stated its intention to "improve the licensing environment for advanced nuclear power reactors to minimize complexity and uncertainty in the regulatory process." The staff noted in its Advanced Reactor Research Plan to the Commission, that a risk-informed regulatory structure applied to license and regulate advanced (future) reactors, regardless of their technology, could enhance the effectiveness, efficiency, and predictability (i.e., stability) of future plant licensing. Therefore, a need was identified for a "Framework" to guide the development of a risk-informed and performance-based approach for future plant licensing for advanced (non-light water) reactors. This NUREG report satisfies that need.

The development of a risk-informed and performance-based regulatory structure for the licensing of diverse reactor designs is a complicated and multi-phase program. Before thoroughly embarking on implementing such an initiative, it is prudent to understand whether such a venture is feasible. The first phase, which is the Framework, is to demonstrate the feasibility of such a concept. The second phase, which involves implementation and would only be pursued upon Commission direction, is comprised of several, iterative steps: resolution of issues, development of draft requirements and regulations, pilots and tests, and rulemaking.

This NUREG report documents one approach to establish the feasibility for development of a risk-informed and performance-based process for the licensing of future nuclear power plants. Part of this documentation is the identification of the programmatic, policy, and technical issues that would need to be addressed by the staff for implementation of such an approach.

This work is intended to assess the feasibility of alternative licensing structures for future designs. It does not represent staff positions with respect to the licensing of current or new plants.

TABLE OF CONTENTS – VOLUME 1
MAIN REPORT

TABLE OF CONTENTS – VOLUME 1 (continued)

TABLE OF CONTENTS – VOLUME 1 (continued)

TABLE OF CONTENTS – VOLUME 1 (continued)

LIST OF FIGURES

TABLE OF CONTENTS – VOLUME 1 (continued)
LIST OF FIGURES

LIST OF TABLES

TABLE OF CONTENTS – VOLUME 2
APPENDICES

TABLE OF CONTENTS – VOLUME 2 (continued)

TABLE OF CONTENTS – VOLUME 2 (continued)

TABLE OF CONTENTS – VOLUME 2 (continued)

LIST OF FIGURES

LIST OF TABLES

TABLE OF CONTENTS – VOLUME 2 (continued)
LIST OF TABLES

TABLE OF CONTENTS – VOLUME 2 (continued)
LIST OF TABLES

EXECUTIVE SUMMARY

The purpose of this NUREG is to establish the feasibility of developing a risk-informed and performance-based regulatory structure for the licensing of future (advanced non-light water reactor (LWR)) nuclear power plants (NPPs). This NUREG documents a "Framework" that provides an approach, scope and criteria that could be used to develop a set of requirements that would serve as an alternative to 10 CFR 50 for licensing future NPPs. This alternative to 10 CFR 50 would have the potential following advantages:

- It would require a broader use of design specific risk information in establishing the licensing basis, thus better focusing the licensing basis, its safety analysis and regulatory oversight on those items most important to safety for that design.

- It would stress the use of performance as the metric for acceptability, thus providing more flexibility to designers to decide on the design factors most appropriate for their design and facilitate the development of an U.S. Nuclear Regulatory Commission (NRC) reactor oversight program that focuses on safety performance.

- It could be written to be applicable to any reactor technology, thus avoiding the time consuming and less predictable process of reviewing non-light-water reactor (LWR) designs against the LWR-oriented 10 CFR 50 regulations, which requires case-by-case decisions (and possible litigation) on what 10 CFR 50 regulations are applicable and not applicable and where new requirements are needed.

This Framework, as such, is not the entire process. It is an initial phase in developing such a regulatory structure. The development of a risk-informed and performance-based regulatory structure for the licensing of diverse reactor designs would be a complicated and multi-phase program. Before thoroughly embarking on implementing such an initiative, it is prudent to understand whether such a venture is feasible. Therefore, the first phase, and major objective of this report, is to demonstrate the feasibility of such a concept, recognizing that for full implementation there would be outstanding programmatic, policy, and technical issues to be resolved. The second phase, which would involve implementation, is comprised of several, iterative steps: resolution of issues, development of draft requirements and regulations, pilots and tests, and rulemaking. Figure ES-1 shows this multi-phased program

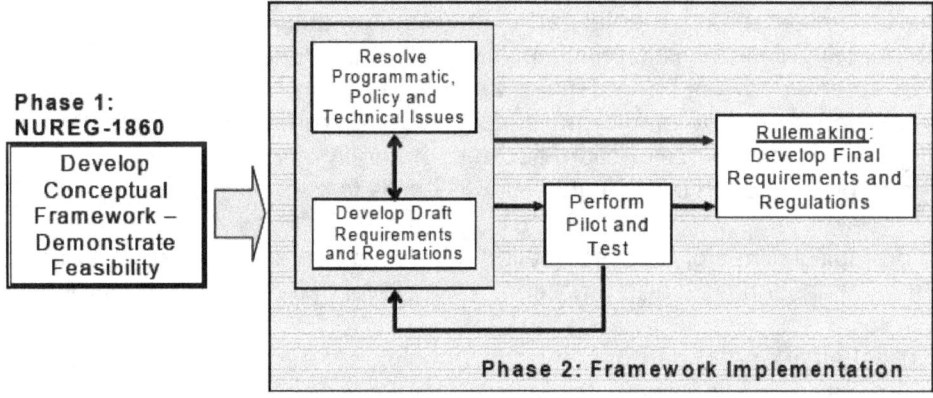

Figure ES-1   Development of Regulatory Structure for
Future Plant Licensing

This first phase, feasibility of a conceptual framework, is the focus and objective of this report. The technical objectives this NUREG is intended to achieve, in establishing feasibility, are:

- be risk-informed
- be performance-based
- incorporate defense-in-depth
- provide flexibility.

Achievement of these objectives should result in a more effective, efficient and stable licensing process for advanced non-LWR designs.

Similar to 10 CFR 50, it covers the design, construction and operation phases of the plant lifecycle up to and including the initial stages of decommissioning (i.e., where spent fuel is still stored on-site). It covers the reactor and support systems. Fuel handling and storage are not addressed, but rather would be considered as part of implementation. The technical basis and process described in this NUREG are directed toward the development of a stand alone set of requirements (containing technical as well as administrative items) that would be compatible and interface with the other existing parts of 10 CFR (e.g., Part 20, 51, 52, 73, 100, etc.) just as 10 CFR 50 is today. The approach taken in developing the technical basis and process is one that integrates deterministic and probabilistic elements and builds upon recent policy decisions by the Commission related to the use of a probabilistic approach and mechanistic radioactive source terms in establishing the licensing basis.

At the highest level, the Framework has been developed from the top down with the safety expectation that future NPPs are to achieve a level of safety at least as good as that defined by the Quantitative Health Objectives (QHOs) in the Commission's 1986 Safety Goal Policy Statement. This approach is consistent with the Commission's 1986 Policy Statement on Advanced Reactors which states that the Commission expects advanced reactor designs will comply with the Commission's Safety Goal Policy Statement, and is discussed further in Chapter 3. Possible criteria are then developed, consistent with the QHOs, that utilize a probabilistic approach for defining the licensing basis (discussed later). Implementation of these criteria would require a design specific probabilistic risk assessment (PRA) and would result in a design specific licensing basis.

Defense-in-depth remains a fundamental part of the requirements development process and has as its purpose applying deterministic principles to account for uncertainties. Defense-in-depth is discussed in Chapter 4 and has been defined as an element in NRC's safety philosophy that is used to address uncertainty by employing successive measures, including safety margins, to prevent or mitigate damage if a malfunction, accident, naturally or intentional caused event occurs. The defense-in-depth approach taken in this NUREG calls for:

- providing multiple lines of defense (called protective strategies) against off-normal events and their consequences which represent a high level defense-in-depth structure; and

- the application of a set of defense-in-depth principles to each protective strategy that result in certain deterministic criteria to account for uncertainties (particularly completeness uncertainties).

The protective strategies, discussed in Chapter 5, address accident prevention and mitigation and consist of the following:

- physical protection (provides protection against intentional acts);

- stable operation (provides measures to reduce the likelihood of challenges to safety systems);

- protective systems (provides highly reliable equipment to respond to challenges to safety);

- barrier integrity (provides isolation features to prevent the release of radioactive material into the environment); and

- protective actions (provides planned activities to mitigate any impacts due to failure of the other strategies).

These protective strategies, in effect, provide for successive lines of defense, each of which would need to be included in the design.

The defense-in-depth principles, discussed in Chapter 4, would require designs to:

- provide measures against intentional as well as inadvertent events;

- provide accident prevention and mitigation capability;

- ensure key safety functions are not dependent upon a single element of design, construction, maintenance or operation;

- ensure uncertainties in equipment and human performance are accounted for and appropriate safety margins provided;

- provide alternative capability to prevent unacceptable releases of radioactive material to the public; and

- be sited at locations that facilitate protection of public health and safety.

As discussed earlier, a set of probabilistic criteria (Chapter 6) have been developed consistent with the Safety Goal QHOs that address:

- overall plant risk and the use of risk-information in design, construction and operations;
- allowable consequences of event sequences versus their frequency;
- selection of event sequences which to be considered in the design;
- safety classification of equipment; and
- security performance standards.

These criteria would also replace the single failure criterion, unless imposed as a defense-in-depth consideration. The approach taken in the Framework continues the practice of ensuring that the allowable consequences of events (defined in Chapter 6) are matched to their frequency such that frequent events are to have very low consequences and less frequent events can have higher consequences. This is expressed in the form of a frequency-consequence (F-C) curve as discussed in Chapter 6. The allowable consequences are based upon existing dose limits, and the associated frequencies are based on guidance contained in International Commission on Radiological Protection 64 and engineering judgment.

Certain event sequences (defined in Chapter 6) from the design specific PRA are chosen for use in establishing plant design parameters for safe operation and equipment safety classification. These events are called licensing basis events (LBEs) and are sequences from the PRA that have to meet stringent acceptance criteria related to the F-C curve and additional deterministic criteria that depend on three broad ranges of accident frequency:

frequent	$\geq$	10^{-2}/yr		
infrequent	$<$	10^{-2}/yr but	$\geq$	10^{-5}/yr
rare	$<$	10^{-5}/yr but	$\geq$	10^{-7}/yr

Chapter 6 provides additional descriptions of the event categories, the LBE selection process , acceptance criteria, analysis guidelines and additional discussion on the safety classification process.

As discussed above, risk assessment would have a more prominent and fundamental role in the licensing process than it does today under 10 CFR 50, since the risk assessment would be an integral part of the design process and licensing analysis. Because of this more prominent use of PRA, the Framework is considered fully risk-informed. Therefore, a high level of confidence would be needed in the results of the risk assessment used to support licensing. In addition, under this risk-informed licensing approach, the risk assessment would need to be maintained up to date over the life of the plant, since it would be an integral part of decision-making with respect to operations (e.g., maintenance, plant configuration control), plant modifications, and maintaining the licensing basis up to date (e.g., assessing the impact of plant operating experience, modifications, etc. on items such as safety classification, LBEs, etc.). Possible guidance on the scope and technical acceptability of the risk assessment needed to support this licensing approach is provided in Chapter 7.

Chapter 8 describes the process for developing potential requirements consistent with the guidance in Chapters 3 through 7. The process for identifying the potential requirements begins with the protective strategies. Each one is examined with respect to what are the various threats or challenges that could cause the strategy to fail. These challenges and threats are identified using a logic tree to perform a "systems analysis" of the strategy to identify potential failures. The defense-in-depth principles are then applied to each protective strategy. Defense-in-depth measures are identified which are incorporated into the potential requirements to help prevent protective strategy failure. This approach forms the process for the selection of "topics." Potential hypothetical requirements are then identified for each topic.

Part of the process involves development of guidance that would be used for actually writing the requirements. This guidance addresses writing the requirements in a performance-based fashion, incorporating lessons learned from past experience, and utilizing existing requirements and guidance, where practical. The guidance also would ensure that the probabilistic process for establishing the licensing basis are incorporated. All of the above are integrated and results in a set of potential requirements which serve to illustrate and establish the feasibility of developing a risk-informed and performance-based licensing approach.

A set of potential requirements is provided in Appendix J. A few examples include the following:

- **Potential Design Requirement #2: Criteria for Selection of the Licensing Basis**

  "Event sequences from the design specific PRA which needs to be considered in the licensing analysis needs to be categorized as follows:

  - frequent $\geq 10^{-2}$/reactor year (ry) (mean frequency)
  - infrequent $< 10^{-2}$/ry but $\geq 10^{-5}$/ry (mean frequency)
  - rare $< 10^{-5}$/ry but $\geq 10^{-7}$/ry (mean frequency)

  Within each of these categories, the applicant/licensee need to designate those sequences of each event type (e.g., loss of coolant accidents, external events, etc.) with the largest consequences as Licensing Basis Events (LBEs) which need to meet the acceptance criteria in Design Requirement #3.

  A postulated LBE for plant siting purposes needs to be selected in accordance with and meet the acceptance criteria in Design Requirement #8."

  (This potential requirement does not have an equivalent in 10 CFR Part 50.)

- **Potential Design Requirement #27: Control Room Design**

  "The main control room needs to be designed with sufficient shielding and atmospheric control to ensure habitability by control room personnel for all accident sequences that have a frequency greater than 10^{-7}/ry (mean value). Habitability needs to encompass assuring the dose to control room operating personnel does not exceed 5 rem for the duration of the accident and that hazardous chemicals are not allowed entry in sufficient concentrations to affect the health and safety of control room personnel.

  The control room needs to have sufficient instrumentation, control and communication capability to allow all safety significant functions to be performed from this location."

  (This potential requirement would be the equivalent to GDC #19.)

- **Potential Design Requirement #29: Reactor Core Flow Blockage and Bypass Prevention**

  "Each reactor design needs to provide measures to prevent bypass and blockage of flow through the reactor core that is sufficient to cause localized fuel damage."

  (This potential requirement does not have an equivalent in 10 CFR Part 50.)

A completeness check was also made by comparing the draft example requirements to other safety requirements documents (e.g., International Atomic Energy Agency (IAEA) Standards, 10 CFR 50). The results of the completeness check are discussed in Chapter 8, and generally concludes that the topics identified are reasonably complete.

In addition, there are a number of programmatic, policy and open technical issues that would need to be resolved if, and when, a decision is made to pursue Framework implementation. These issues are described in Appendix C. The programmatic issue addresses the manner in which, if decided by the Commission, the Framework should be implemented (e.g., technology-neutral versus technology-specific, rule-making versus design-specific). The policy issues, for example, include such items as level of safety (e.g., acceptability of using the QHOs as the level of safety new plants are to achieve); and integrated risk (e.g., apply the QHOs on a per reactor or per site basis). The technical issues, for example, would involve such items as use of a complementary cumulative distribution function as an additional risk criterion; assessment of environmental protection; development of risk objectives subsidiary to the QHOs addressing accident prevention and mitigation; and development of risk-importance measures for non-LWRs and guidance for their use. However, the fact that a number of open items remain does not detract from the validity of the technical information contained in this document.

Finally, Chapter 9 discusses the conclusions and steps needed if, and when, the requirements resulting from application of the Framework would be implemented and used in plant licensing.

In summary, this NUREG has met the objectives and established the feasibility of developing a risk-informed and performance-based approach for future plant licensing. This conclusion is based upon the successful development of risk criteria that would be implemented using design-specific risk information, integration of probabilistic and deterministic (e.g., defense-in-depth) elements, demonstration of the LBE selection and safety classification process, development of potential requirements and the results from the check against other requirements documents.

In addition to the resolution of programmatic policy and technical issues described above, the following steps would also need to be taken to fully implement the Framework:

- completion of requirements development,
- development of implementing guidance,
- pilot testing on an actual reactor design,
- reactor oversight program development, and
- rule-making, if necessary.

This NUREG report documents one approach to establish the feasibility for development of a risk-informed and performance-based process for the licensing of future nuclear power plants. Part of this documentation is the identification of the programmatic, policy, and technical issues that would need to be addressed by the staff for implementation of such an approach.

This work is intended to assess the feasibility of alternative licensing structures for future designs. It does not represent staff positions with respect to the licensing of current or new plants.

ACKNOWLEDGMENTS

This report documents the feasibility of developing a risk-informed and performance-based regulatory structure for the licensing of future nuclear power plants. It documents a "Framework" that provides an approach, scope and criteria that could be used to develop a set of requirements that would serve as an alternative to 10 CFR 50 for licensing future NPPs.

Overall management and technical leadership of the project was provided by: Mary Drouin

The principal technical authors[1] of this report, by alphabetic order, were:

Dennis Bley (Buttonwood Consulting)   John Lehner (BNL)
Mary Drouin (NRC)   Bruce Mrowca (ISL)
Tom King (ISL)   Vinod Mubayi (BNL)
Jeffrey LaChance (SNL)

The authors would like to acknowledge the technical contributions made by numerous other NRC staff and other individuals that contributed to the development of the Framework. They include, by alphabetic order:

Charles Ader   John Monninger
Benjamin Beasley   Bruce Musico
Les Cupidon   Gareth Parry
Farouk Eltawila   Trevor Pratt (BNL)
Mirela Gavrilas   Joshua Reinert (ISL)
Eric Haskins (independent consultant)   Stuart Rubin
Alan Kuritzky   Martin Stutzke
Inn Kim (formerly of ISL)   Ashok Thadani

The authors would like to acknowledge the independent peer reviewers. Their comments contributed considerably to the improved clarity and technical quality of this document.

Christopher Grimes (retired NRC, independent consultant)
Thomas Murley (retired NRC, independent consultant)

Finally, the authors would like to acknowledge the numerous NRC staff and other individuals who contributed in the preparation, assembly, graphics, etc. of this document. They include, by alphabetic order:

Colleen Amoruso (ISL)   Paul Kleene
Danielle Burnette   Tammy Pfiester (ISL)
Cheryl Conrad (BNL)   Lauren Killian
Jean Frejka (BNL)   Katherine Orta (BNL)
Michelle Gonzalez   Brian Wagner

---

[1]ISL – Information Systems Laboratory, SNL – Sandia National Laboratories, BNL – Brookhaven National Laboratory

ACRONYMS AND ABBREVIATIONS

ACR	Advanced CANDU Reactor
ACRS	Advisory Committee on Reactor Safeguards
AEC	Atomic Energy Commission
ALARA	As Low As Reasonably Achievable
ALWR	Advanced Light Water Reactor
ANPR	Advance Notice of Proposed Rulemaking
ANS	American Nuclear Society
AO	Abnormal Occurrence
AOO	Anticipated Operational Occurrences
ASME	American Society of Mechanical Engineer
ATWS	Anticipated Transient Without Scram
BDBT	Beyond Design Basis Threat
CCDF	Complementary Cumulative Distribution Function
CCFP	Conditional Containment Failure Probability
CDF	Core Damage Frequency
CFR	Code of Federal Regulations
CLB	Current Licensing Basis
CO	Carbon Monoxide
CO_2	Carbon Dioxide
CP	Construction Permit
CPEF	Conditional Probability of Early Fatality
CPLF	Conditional Probability of Latent Fatality
DCF	Dose Conversion Factor
DBA	Design Basis Accidents
DBT	Design Basis Threat
EAB	Exclusion Area Boundary
ECC	Emergency Core Cooling
ECI	Emergency Coolant Injection
EF	Early Fatality
EIS	Environmental Impact Statement
EP	Emergency Preparedness
EPA	Environmental Protection Agency
EXF[1]	1 rem Exceedance Frequency
F-C	Frequency Consequence
FSAR	Final Safety Analysis Report
F-V	Fussell-Vesely
GDC	General Design Criteria
GFR	Gas-cooled Fast Reactor
GPRA	Government Performance and Results Act
GTCC	Greater than Class C
HAZOP	Hazard and Operability Analysis
HEU	Highly Enriched Uranium
HLR	High Level Requirement
HLW	High Level Waste
HSE	Health and Safety Executive
HTGR	High Temperature Gas-cooled Reactor
IAEA	International Atomic Energy Agency
I&C	Instrumentation and Control
ICRP	International Commission on Radiation Protection
IEEE	Institute of Electrical and Electronics Engineers
IER	Individual Early Risk
ILR	Individual Late Risk

ACRONYMS AND ABBREVIATIONS (continued)

IM	Importance Measure or Measures
ISFSI	Independent Spent Fuel Storage Installation
ISGTR	Induced Steam Generator Tube Rupture
LBE	Licensing Basis Events
LERF	Large Early Release Frequency
LEU	Low Enriched Uranium
LF	Latent Fatality
LFR	Lead-cooled Fast Reactor
LLRF	Large Late Release Frequency
LMR	Liquid Metal-cooled Reactor
LPZ	Low Population Zone
LOCA	Loss of Coolant Accidents
LPZ	Low Population Zone
LTC	Long Term Cooling
LWR	Light Water Reactors
MOX	Mixed Oxide
MSR	Molten Salt Reactor
MW_e	Mega-watt Electric
NDE	Non Destructive Examination
NEPA	National Environmental Policy Act
NERI	Nuclear Energy Research Initiative
NGNP	Next Generation Nuclear Plant
NPP	Nuclear Power Plant
NRC	U.S. Nuclear Regulatory Commission
OL	Operating License
PAG	Protective Action Guidelines
PBMR	Pebble Bed Modular Reactor
PCT	Peak Cladding Temperature
PRA	Probabilistic risk assessments
PSAR	Preliminary Safety Analysis Report
Pu	Plutonium
QA	Quality Assurance
QC	Quality Control
QUO	Quantitative Health Objectives
RAW	Risk Achievement Worth
RCCS	Reactor Cavity Cooling System
RCS	Reactor Coolant System
RI/PB	Risk-Informed and Performance-Based
ROP	Reactor Oversight Process
SAMDA	Severe Accident Mitigation Design Alternative
SAR	Safety Analysis Report
SBO	Station Blackout
SCWR	Super Critical Water Reactor
SFR	Sodium-cooled Fast Reactor
SGTR	Steam Generator Tube Rupture
SNM	Special Nuclear Material
SRM	Staff Requirements Memorandum
SSC	Systems, Structures and Components
TEDE	Total Effective Dose Equivalent
TP	Total Population
U.K.	United Kingdom
VHTR	Very High Temperature Reactor

1. INTRODUCTION

1.1 Background

The U.S. Nuclear Regulatory Commission, in its Policy Statement on Regulation of Advanced Nuclear Power Plants, [NRC 1994] stated its intention to "improve the licensing environment for advanced nuclear power reactors to minimize complexity and uncertainty in the regulatory process." The staff noted in its Advanced Reactor Research Plan [NRC 2002] to the Commission that a risk-informed regulatory structure applied to license and regulate advanced (future) reactors, regardless of their technology, could enhance the effectiveness, efficiency, and predictability (i.e., stability) of future plant licensing. Therefore, a need was identified for a "Framework" to guide the developing of a risk-informed and performance-based approach for future plant licensing.

This need, to develop a risk-informed, performance-based framework for establishing technology-neutral or technology-specific requirements, for future reactors is based on the following considerations:

- The regulatory structure for current light water reactors (LWRs) has evolved over five decades. Most of this evolution occurred without the benefit of insights from probabilistic risk assessments (PRAs) and severe accident research. The use of risk metrics in evaluating safety focuses attention on those areas where risk is most likely. Therefore, it is expected that the regulations for future reactors will be risk-informed. It is anticipated that deterministic and probabilistic criteria and results will be used in developing the regulations governing these reactors. Consequently, a structured approach toward a regulatory structure for future reactors that incorporates probabilistic and deterministic insights will help ensure the safety of these reactors by focusing the regulations on where the risk is most likely. At the same time it needs to maintain basic safety principles, such as defense-in-depth and safety margin. Therefore, it is expected that future applicants will rely on PRAs as an integral part of their license applications. Hence guidance and criteria on the use of PRA results and insights will be an important aspect of the licensing process.

- In 1993, Congress passed a law called the "Government Performance and Results Act" (GPRA) [US 1993]. One objective of that law is to "improve Federal program effectiveness and public accountability by promoting a new focus on results, service, quality, and customer satisfaction." In response to the GPRA, Federal agencies, including the U.S. Nuclear Regulatory Commission (NRC), developed strategies and plans for achieving that objective. The use of performance measures provides flexibility to designers in emphasizing outcomes rather than prescriptive methods of achieving them. The NRC, in its strategic plan, committed to establish, where appropriate, performance-based regulations and to use performance-based regulation to minimize unnecessarily prescriptive requirements.

- In 1995, the Commission issued a policy statement on the use of probabilistic risk assessment (PRA) [NRC 1995] methods in nuclear regulatory activities. One purpose of the policy statement was to ensure that the many potential applications of PRA were implemented in a consistent and predictable manner that would promote regulatory stability and efficiency. The policy statement directed that the use of PRA technology should be increased in all regulatory matters to the extent supported by the state-of-the-art in PRA methods and data, and in a manner that complements the U.S. Nuclear Regulatory Commission's (NRC's) deterministic approach and supports the NRC's traditional defense-in-depth philosophy.

1. Introduction

- On March 11, 1999, the NRC stated in Yellow Announcement #019 [NRC 1999] that "The Commission has issued a white paper that defines the terms and Commission expectations regarding risk-informed and performance-based regulation." The Commission in the white paper stated that: "The Commission is advocating certain changes to the development and implementation of its regulations through the use of risk-informed, and ultimately performance-based, approaches. The PRA Policy Statement formalized the Commission's commitment to risk-informed regulation through the expanded use of PRA..." The paper further noted that "a performance-based approach focuses on a licensee's actual performance results (i.e., desired outcomes), rather than on products (i.e., outputs). In the broadest sense, the desired outcome of a performance-based approach to regulatory oversight will be to focus more attention and NRC resources on those licensees whose performance is declining or less than satisfactory."

- While the NRC has over 30 years experience with licensing and regulating nuclear power plants, this experience (as reflected in regulations, regulatory guidance, policies and practices) has been focused on current light-water-cooled reactors (LWRs) and may have limited applicability to future reactors. The design and operational issues associated with the future reactors may be distinctly different from current LWR issues. The current set of regulations does not necessarily address safety concerns that may be posed by new designs, and may contain specific requirements that do not pertain to new designs.

- The provision of a Framework, which can be developed for diverse reactor designs using important probabilistic and deterministic criteria governing risk and performance, will facilitate developing a consistent, stable, and predictable set of requirements that are both risk-informed and performance-based. These requirements may be either technology-neutral (can be applied to any reactor design in conjunction with technology-specific regulatory guides), or technology-specific, i.e., focused on particular reactor technologies (e.g., high-temperature gas reactors).

The NRC's LWR experience, especially the recent efforts to risk-inform the regulations, has shown the potential value of a top-down approach to developing a regulatory structure for a new generation of reactors. Such an approach could facilitate implementing risk-informed, performance-based regulation, as well as ensure a greater degree of coherence among the resulting regulations for future reactors than found among current regulations.

In addition to using the benefits of PRA, the development of a risk-informed and performance-based structure for future plant licensing has several potential advantages over continuing to use Part 50 of Title 10 of the Code of Federal Regulations (10 CFR Part 50) licensing process for designs substantially different from a current generation LWRs. While the current Part 50 requirements are used to the extent feasible in developing the alternative, the use of a technology-neutral approach can potentially provide greater effectiveness, efficiency, stability and predictability than continuing to use the 10 CFR Part 50 process. These points are further discussed below.

- **Effectiveness**: To be effective, the regulations should focus on those items most important to the protection of public health and safety. Since reactor designs and technologies can be different, the regulations also need to be able to be applied to such variations without any loss of safety focus. Currently, 10 CFR 50 focuses on LWR safety, but does not vary with different LWR designs, except in a few selected areas (e.g., external events). Accordingly, a technology-neutral set of requirements that accomplishes the above would potentially provide improved effectiveness and flexibility without a loss of safety.

- **Efficiency:** When 10 CFR Part 50 is used to license a reactor design that is substantially different than a current generation LWR, the regulations need to be reviewed for applicability to that design. In the review, determinations need to be made on which regulations apply, which do not, which require modification, and what additional requirements are needed to address the unique aspects of the design under review. Once these determinations are made, exemptions need to be processed to formally document the rules that do not apply, and the Commission may need to approve any new requirements (as was done in the certification of the advanced light water reactors [ALWRs]). The results of this process are also subject to challenge through the intervention and hearing process. This entire process needs to be done for each design reviewed using 10 CFR Part 50. Repeating this process for each new design is inefficient. A technology-neutral licensing process that applies regardless of reactor design will eliminate the case-by-case review process.

- **Stability/Predictability:** Putting each reactor design through the licensing process described above does not lead to stability in licensing. With case-by-case reviews and intervention, similar issues may have different resolutions. This situation can occur due to different staff, Commission, or public involvement. This licensing process has large uncertainties in both outcome and duration. A licensing process derived from a technology-neutral framework establishes a level playing field based on risk criteria and fundamental safety principles like defense-in-depth and safety margin and has acceptance criteria applicable to all reactor designs. This approach will reduce the uncertainties in the outcome and duration of the licensing process because acceptance criteria would be stable. Having a stable set of requirements that are based on and derived from technology-neutral criteria and principles will promote predictability by having a common set of goals, criteria, and methods that the NRC, applicants, and licensees use. Predictability is an important factor in any decision to pursue the licensing of a nuclear power plant.

The development of a licensing process based on a top-down framework (that can be applied to any reactor design) that is an alternative to the current Part 50 process will help ensure that a systematic approach is used to develop the regulations for the design, construction, and operation of future reactors. This new licensing process could potentially ensure a greater degree of uniformity, consistency, and defensibility in developing the requirements, particularly when addressing the unique design and operational aspects of future reactors.

1.2   Objectives

The development of a risk-informed and performance-based regulatory structure for the licensing of diverse reactor designs is a complicated and multi-phase program. Before thoroughly embarking on implementing such an initiative, it is prudent to understand whether such a venture is feasible. Therefore, the first phase, and major objective of this report, is to demonstrate the feasibility of such a concept, recognizing that for full implementation there will be outstanding programmatic, policy, and technical issues to be resolved. The second phase, which involves implementation, is comprised of several, iterative steps: resolution of issues, development of draft requirements and regulations, pilots and tests, and rulemaking.

This first phase, feasibility of a conceptual framework, is the focus and objective of this report. Figure 1-1 shows this multi-phased program.

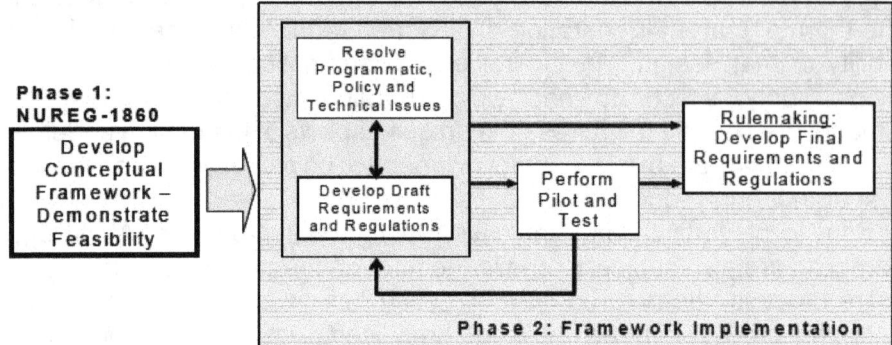

Figure 1-1   Development of Regulatory Structure for
Future Plant Licensing

The developed concept or approach (in the form of guidelines and criteria) would serve as the initial technical basis for implementing a risk-informed, performance-based framework for licensing future commercial nuclear power plants.

In addition to feasibility, the framework should also achieve the following objectives:

•   ***Risk-informed*** — Ensure that risk information and risk insights are integrated into the decision making process such that there is a blended approach using both probabilistic and deterministic information.

•   ***Performance-based*** — When implemented, the guidance and criteria produce a set of safety requirements that are based on plant performance, and do not use prescriptive means for achieving its goals.

•   ***Defense-in-depth*** — Defense-in-depth is an integral part of the framework such that uncertainties are accounted for in the requirements for design, construction, and operation.

•   ***Flexible*** — The framework should allow the licensing process to support reactors of diverse designs and be developed in such a manner that, as new information and knowledge, is gained, changes to the regulatory structure can be implemented effectively and efficiently.

The above provide the set of fundamental objectives for the framework. However, while it may be possible to develop and demonstrate feasibility of a framework that meets these objectives, there are sub-objectives that are also desirable so that the framework will yield a set of cohesive and integrated requirements that are risk-informed, performance-based for reactors of diverse design.

These include the following:

•   Consistency with Commission Policies – the framework should implement and be consistent with Commission policy.

•   Compatible with other parts of 10 CFR – the framework should interface with the other parts of 10 CFR and minimize any conforming changes.

- Compatible with Licensing Process – the framework should allow either a two-step licensing process (i.e., construction permit/operating license) or a one-step (combined operating license) licensing process, similar to the current 10 CFR Part 50. It should also include a provision for exemptions in case an applicant wishes to propose an alternative approach to one or more requirements.

- Builds upon existing regulatory experience – the framework should acknowledge experience and take advantage of lessons learned from this experience.

The above sub-objectives need to be met to ensure that the main objectives are accomplished in a fashion that results in an improved licensing process; that is, an effective, efficient, and stable licensing process. Each of the above sub-objectives is applicable to each of the main objectives and should be considered when assessing whether the main objectives have been met.

1.3   Scope

The risk-informed, performance-based framework developed in this report can be applied to all future plants. The regulations that derive from this framework could apply to all types of reactor designs, including gas-cooled, liquid metal, and heavy and light-water-moderated reactors. This applicability will be accomplished either by (1) having the regulatory requirements specified at a high (technology-neutral) level with accompanying technology-specific regulatory guides, or (2) developing technology-specific requirements for particular designs based on the criteria and guidance offered in the framework. Appendix A provides a summary of the safety characteristics of the various reactor technologies.

The framework addresses risks from all sources of radioactivity that are present at the plant except for fuel storage, handling, and rad waste. These include reactor full-power, low-power and shut-down operation, and the risks from both internal and external events. Therefore, it includes seismic, fire and (internal and external) flood risks, and risk from high winds and tornados. Issues related to security are also considered. Risks from other sources that are an integral part of the licensing process, e.g., liquid sodium for liquid metal reactors, are also included in the scope of the framework.

The framework covers design, construction, and operation. Operation includes both normal operation as well as off-normal events, ranging from anticipated occurrences to rare but credible events, for which accident management capabilities may be needed.

The framework is intended to provide guidance on the structure and key elements which will be used to develop the risk-informed, performance-based regulations that may be technology-neutral or technology-specific. In effect, the framework provides guidance on key programmatic, policy, and technical issues and the scope of the technology-neutral or technology-specific requirements. However, many of the details will only be developed if, and when, implementation of the framework is pursued.

1.4   Framework

In developing the "blueprint" to be used for the framework, several fundamental questions needed to be explored. That is, the overall conceptual framework proposed in this report, is developed by examining each of the identified questions. Each question translates into an element defining the conceptual framework. The derivation of the framework is briefly described below. Chapter 2 gives an overview, with detailed discussions of each element in the subsequent chapters.

Six questions were identified in developing the conceptual framework. They include the following:

(1)   What approach should be taken to develop the framework?
(2)   What should be the starting point and structure of the framework?
(3)   What should be the role of defense-in-depth?
(4)   What overall process should be used to identify potential requirements?
(5)   What should be the balance between deterministic and probabilistic elements?
(6)   What process should be used to integrate the framework elements?

These questions were explored within the bounds set by the objectives defined for the framework.

Approach

What basic approach should be taken? Should the framework be based on the current Part 50; that is, on revising and modifying Part 50 to meet the objectives? Or, should the framework be developed starting with a "clean sheet of paper?"

Starting with the current Part 50 and its GDCs was determined to be the less desirable approach. Since Part 50 (and the GDCs) is LWR-focused, prescriptive, and deterministically based, it was more prudent to start fresh without any conceived "biases." However, checking the developed framework against Part 50 (and the GDCs) would be important. Figure 1-2 shows this first step in establishing the framework structure.

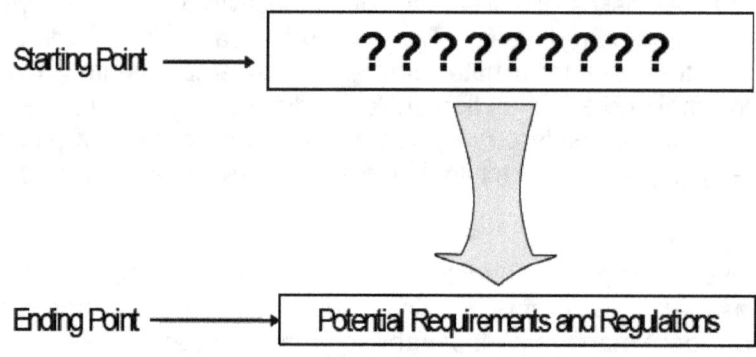

Figure 1-2   Starting Point in Development
                    of Framework Structure

Starting Point and Structure

What should be the starting point of the framework? What should be the structure of the framework? Should the structure be based on a high-level, top-down approach starting with some basic principles and goals? Or, should the framework be more of a bottom-up approach starting with fundamental design criteria?

An optimum approach is an hierarchal top-down structure. It would start at the highest level in defining the basic goals and criteria: Atomic Energy Act [US 1954] which provides the mission to ensure that commercial nuclear power plants are operated in a manner that provides adequate protection of public health and safety and is consistent with the common defense and security. The other top level criteria address the expectations for safety, security, and preparedness. Figure 1-3 shows this simplified "framework."

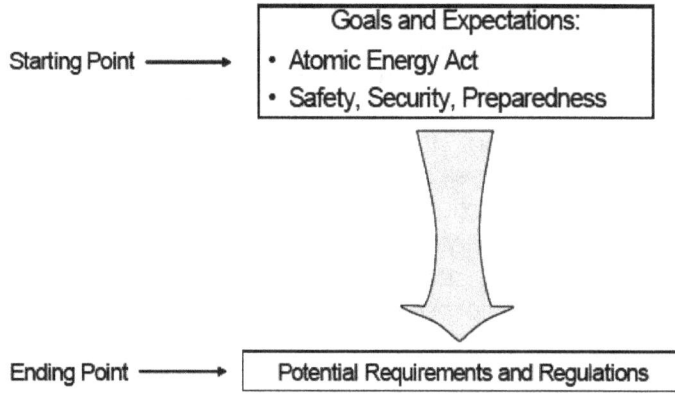

Figure 1-3   Hierarchal Approach in Development of Framework Structure

Defense-in-Depth

What should be the role of defense-in-depth? How should defense-in-depth be factored into the framework? What should be its purpose? How should it relate to uncertainties?

The core of NRC's safety philosophy has always been the concept of defense-in-depth, and it remains basic to this framework. As used in the Framework, the ultimate purpose of defense-in-depth is to compensate for uncertainty, and in this regard it is an integral element of the framework. Principles for defense-in-depth are defined to guide the development of requirements. Figure 1-4 shows this progression of the "framework."

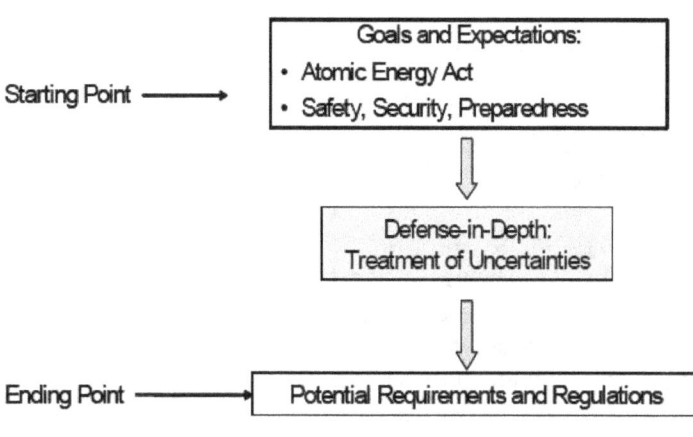

Figure 1-4   Integration of Defense-in-depth in Development of Framework

Process

What overall process should be used to identify potential requirements for design, construction, and operation? Should the process to identify the specific requirements and regulations start with design considerations? Or, should the process start with some other consideration?

1. Introduction

The process chosen to initially identify and define the requirements and regulations needs to start with safety, security, and preparedness expectations, which if met, ensure protection of the public health and safety. Safety, security and preparedness expectations have been defined and implemented in the form of protective strategies that, if met, will ensure the protection of the public health and safety with a high degree of confidence. Figure 1-5 shows this progression of developing the framework.

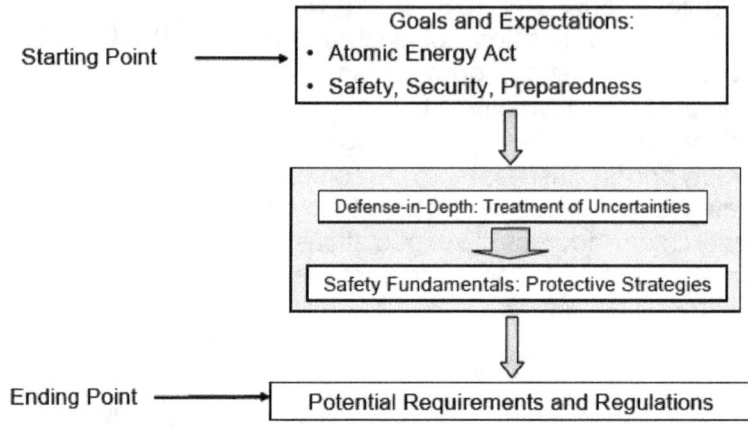

Figure 1-5   Process Used in Development of Framework Structure

Probabilistic versus Deterministic

What should be the balance between deterministic and probabilistic? Should a deterministic set of requirements be defined first and then refined with risk insights, or should a set be defined using both deterministic and probabilistic criteria in an integrated fashion?

In the current Part 50, the licensing basis is established with a deterministic approach and an LWR focus. Consequently, a stylized set of accidents to be considered are not necessarily risk significant or design applicable. In the Framework, a probabilistic process is used that integrates deterministic criteria (based on plant-specific considerations) to establish the potential requirements. Figure 1-6 shows this element in the progression of the framework.

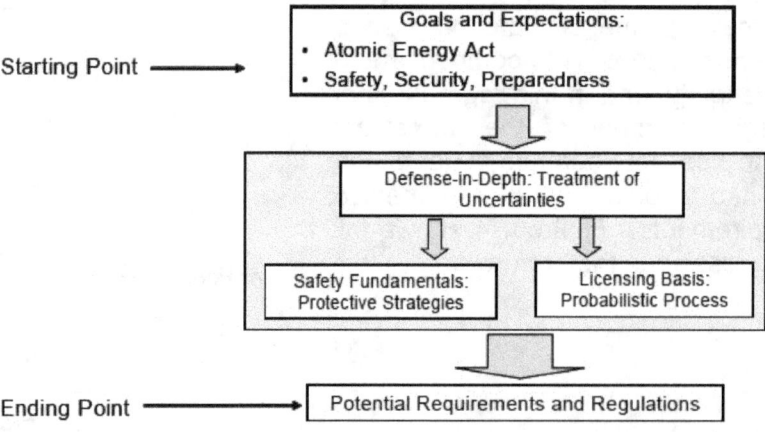

Figure 1-6   Probabilistic Process Used in Development of Framework Structure

Integrated Framework

How should the framework be integrated? Should each of the above questions be treated and addressed separately and independently? Or, should they be integrated to achieve a cohesive set of requirements and regulations?

At this step in the process, the various "framework elements" are brought together and examined. The various challenges or threats that could preclude the element (e.g., protective strategy or defense-in-depth) from being achieved are identified and are used to identify "topics." These topics are the basis for formulating the potential requirements needed to ensure protection of the public health and safety. Figure 1-7 shows this element in the progression of the framework.

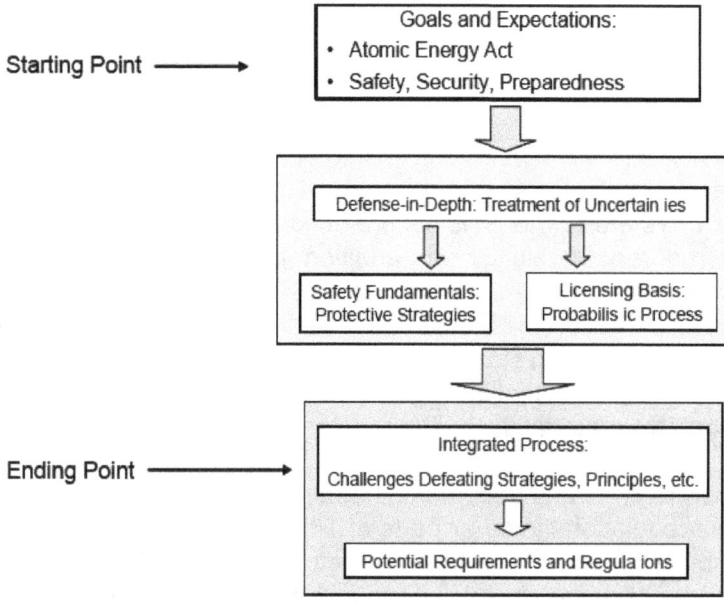

Figure 1-7   Overall Framework Structure

1.5   Relationship to Current Licensing Process

The purpose of this document is to develop a conceptual framework that could serve as the technical basis to support developing a technology-neutral, risk-informed, and performance-based process for the licensing of new nuclear power plants (NPP). As such, it documents an approach that can be used to create a 'level playing field' for all future reactor technologies in terms of the safety criteria to be met. The approach is intended to be used by the NRC staff to develop, for example, a set of regulations that would serve as an alternative to 10 CFR 50 for licensing future NPPs. The regulations developed from the approach could still be used in conjunction with 10 CFR 52 for carrying out the licensing process, i.e., obtaining a combined operating license or design certification.

A key difference in the approach from the current Part 50 approach is the combination of deterministic and probabilistic criteria to establish the plant's safety. In the current Part 50/52 licensing approach, the deterministic calculations carried out for the licensing basis events, i.e., the design basis accidents (DBAs) and, separately, for the probabilistic risk analysis (PRA) are important components of the safety analyses, but there is no direct link between these two components. The approach in this document links the PRA analysis with the licensing basis event selection, design criteria, and SSC selection and treatment.

1.6   Relationship to Code of Federal Regulations

In establishing a risk-informed and performance-based approach to develop a regulatory structure for the licensing of future reactors, it is necessary to review the relationship of the current 10 CFR Part 50 (and Part 52) to the entire set of regulations governing the nuclear fuel cycle. This review is necessary because the approach developed in the framework needs to interface with the other regulations.

These regulations extend from Part 40 that cover the licensing of source material, through Part 70 that covers the licensing of various operations, leading to the fabrication of fuel assemblies; Parts 72 and 63 cover the licensing of reactor spent fuel storage either at the reactor site or in an independent spent fuel storage installation and final disposal in the high level waste repository; Part 73 refers to the physical protection licensing aspects of plants and materials; and Part 100 governs reactor siting. In addition to the regulations that govern individual steps in the manufacture, use, and disposal of fuel, there are cross-cutting regulations that affect every step of the overall fuel cycle. These cross-cutting regulations, for example, include Part 20 that deals with radiation protection standards for the public, the workers and the environment; Part 51 that covers environmental protection regulations; and Part 71 that involves the safe and secure transport of radioactive material, including reactor fuel.

Appendix B, of this document, contains a review of each of the links of the regulations in Part 50 to the regulations in other parts of Title 10 of the Code of Federal Regulations (CFR). Hence, in establishing a new regulatory structure for Part 50, the links to other parts of 10 CFR and the contents of those links will need to be reviewed to ensure they properly interface. No changes (other than conforming changes) would be needed in other parts of 10 CFR to implement the framework. As such, it is important to recognize where 10 CFR 50 interfaces with the other parts of 10 CFR so that these same interfaces can be maintained.

Figure 1-8 shows how the framework relates to 10 CFR. As mentioned previously, it is only providing a risk-informed and performance-based alternative to 10 CFR 50. All other regulations associated with licensing of NPPs or the nuclear fuel cycle would remain unchanged, except for conforming changes. As shown in Figure 1-8, the interfaces are associated with reactor-specific regulations, cross-cutting regulations (i.e., those regulations affecting more than reactor licensing) and fuel-cycle regulations. Accordingly, in developing the framework it is important to ensure that the technical bases, criteria, and structure preserve these interfaces such that there will be a seamless transition to other parts of 10 CFR.

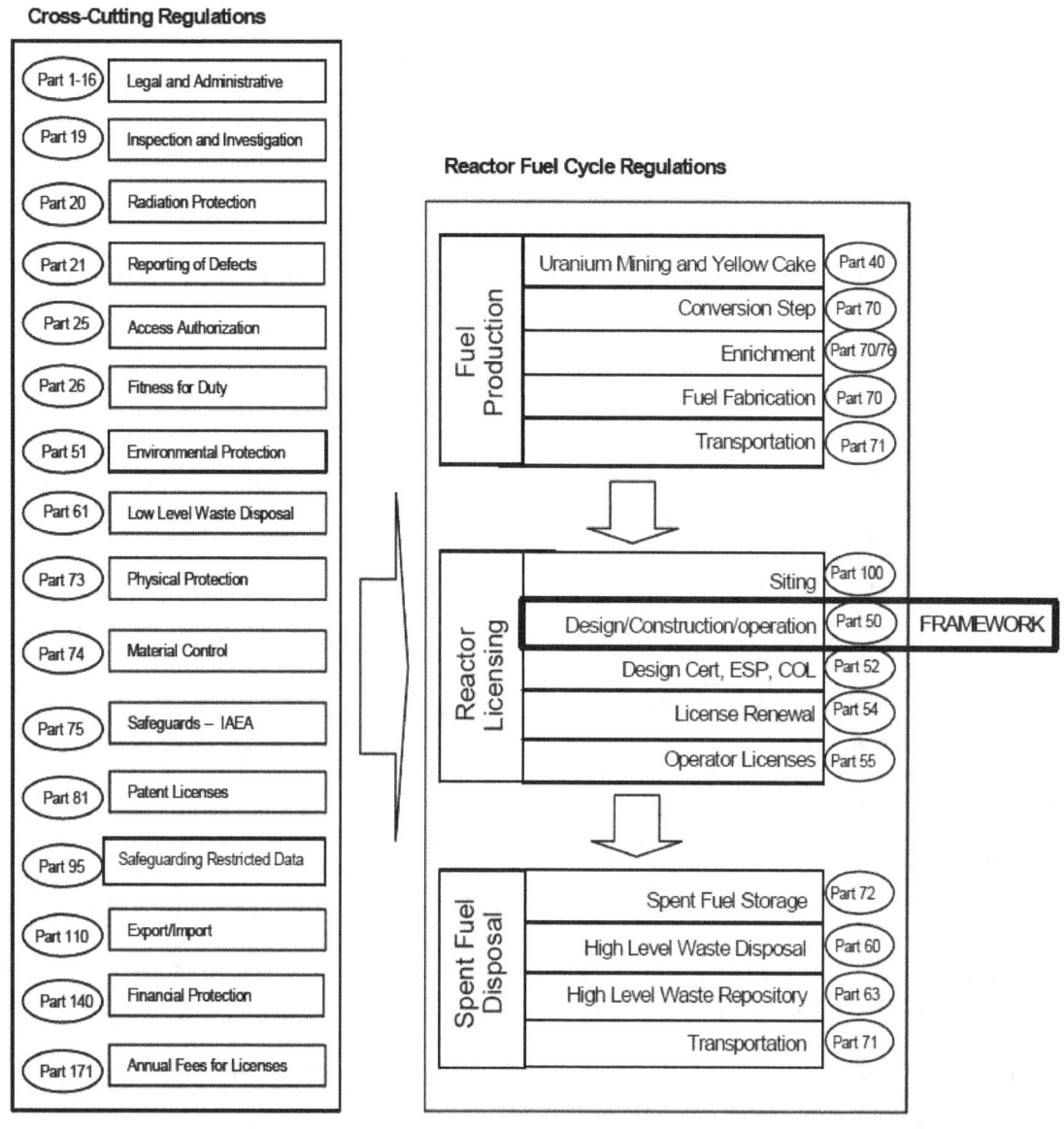

Figure 1-8   Relationship of Framework to Title 10 of Code of Federal Regulations

1.7   Report Organization

The report is organized into 9 Chapters a.

nd 12 Appendices. Following the Introduction, Chapter 2 presents an overview of the framework. Chapters 3 through 8 contain details of the various elements of the framework, and Chapter 9 provides conclusions and discussion on the feasibility and implementation of the framework. Appendices A through K contain additional details of various topics referred to in the chapters, and Appendix L summarizes the various stakeholders comments received on the framework.

1. Introduction

More specifically, the various chapters and appendices address the following:

Chapter 2:   Detailed overview of the framework describing each of the elements.

Chapter 3:   Detailed discussion of high level criteria – safety, security and preparedness expectations to be met.

Chapter 4:   Detailed discussion of defense-in-depth and the treatment of uncertainties.

Chapter 5:   Detailed discussion of the protective strategies defining the safety fundamentals.

Chapter 6:   Detailed discussion of the probabilistic criteria used to establish the licensing basis.

Chapter 7:   Detailed discussion of the technical acceptability needed for the PRA.

Chapter 8:   Detailed discussion of process used to integrate the various elements to identify potential requirements.

Chapter 9:   Detailed discussion on the feasibility and implementation of the framework.

Appendix A: Overview of the safety characteristics of selected future reactors.

Appendix B: Detailed discussion of the relationship of 10 CFR 50 to other parts of 10 CFR and vice versa.

Appendix C: Detailed discussion of the programmatic, policy and technical issues to be addressed for implementing the framework.

Appendix D: Detailed discussion of the derivation of the risk surrogates for LWRs.

Appendix E: Detailed discussion of the results of a test case of the LBE selection and safety classification process.

Appendix F: Detailed discussion of the PRA technical acceptability criteria.

Appendix G: Detailed discussion of the implementation of the process to integrate the framework elements and identify potential requirements.

Appendix H: Detailed discussion of the applicability of 10 CFR 50 requirements to the framework.

Appendix I: Guidance for formulating performance-based requirements.

Appendix J: Detailed list of example draft requirements using the framework.

Appendix K: Detailed discussion of completeness checks performed against the framework.

Appendix L: Detailed summary of the stakeholder comments received and staff response to the comments.

1.8   References

[NRC 1994]   U.S. Nuclear Regulatory Commission, Commission Policy Statement on the Regulation of Advanced Nuclear Power Plants, 59 FR 35461, July 1994.

[NRC 1995]   U.S. Nuclear Regulatory Commission, "Final Policy Statement 'Use of Probabilistic Risk Assessment (PRA) Methods in Nuclear Regulatory Activities'," Washington, DC, 60 FR 42622, August 1995.

[NRC 1999]   U.S. Nuclear Regulatory Commission, Yellow Announcement #019, "Commission Issuance of White Paper on Risk-Informed and Performance-Based Regulation," Shirley Ann Jackson, March 11, 1999.

[NRC 2002]   U.S. Nuclear Regulatory Commission, Advanced Reactor Research Plan, Revision 1, ML021760135, June 2002.

[US 1954]   Public Law 83-703, The Atomic Energy Act of 1954, as Amended, August 1954.

[US 1993]   Public Law 103-62, Government Performance Results Act of 1993, August 1993.

2. FRAMEWORK OVERVIEW

2.1 Introduction

The purpose of this chapter is to provide an overview of the conceptual framework developed for future plant licensing. As noted in Chapter 1, the framework was developed based on a series of questions designed to address the objectives (e.g., risk-informed) and sub-objectives (e.g., consistency with U.S. Nuclear Regulatory Commission (NRC) policy). The questions were explored, and the resolution forms the basic elements of the framework. These questions (or elements) are described below. The details of each element are explored and discussed in detail in the subsequent chapters. The elements include the following, which are also shown in Figure 2-1.

- Element 1: Goals and Expectations
- Element 2: Defense-In-Depth
- Element 3: Safety Fundamentals
- Element 4: Licensing Basis
- Element 5: Integrated Process

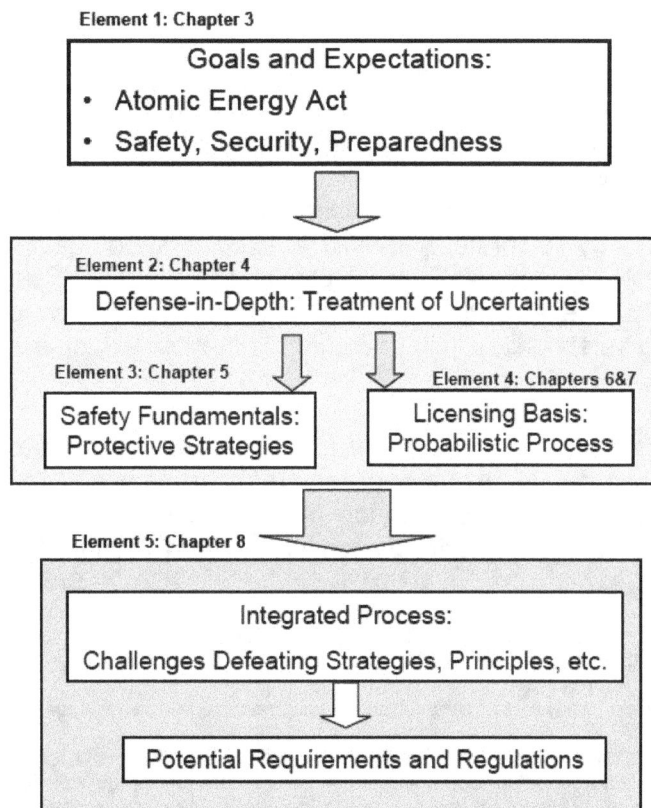

Figure 2-1   Elements of the Framework

2.2   Element 1: Goals and Expectations

The first element in developing the framework is establishing its structure and where it should start. For example, what should be the structure of the framework? Should the structure be based on a high-level, top-down approach starting with some basic principles, goals, and high-level criteria? Or, should the framework be more of a bottom-up approach, starting with fundamental design criteria? What expectations should the framework meet?

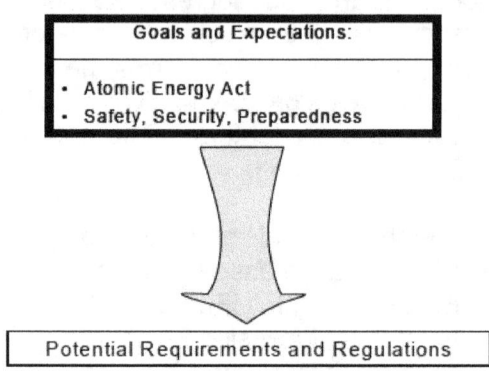

The framework was developed following a top-down approach, as shown in Figure 2-2. It is built upon the traditional NRC safety mission, beginning with the Atomic Energy Act [US 1954] and encompassing a set of safety, security, and preparedness expectations developed from Commission statements.

Figure 2-2   Hierarchal Approach in Development of Framework

The basis for nuclear reactor regulation originates with the Atomic Energy Act of 1954[2] and the statutes that amended it, which indicate that the mission of the NRC, and the Atomic Energy Commission before it, is to ensure that commercial nuclear power plants (NPP) are operated in a manner that provides adequate protection of public health and safety and is consistent with the common defense and security, i.e., protects against radiological sabotage and the theft or diversion of special nuclear materials. The Atomic Energy Act sets the overall NRC safety mission to protect public health and safety. The amending statutes and the broad body of USNRC regulation implement an underlying safety philosophy that controls the risk to workers, off-site populations, and surrounding lands (i.e., environment).

The framework integrates the NRC's expectations for safety, security, and preparedness to achieve the desired overall level of safety. The approach requires that safety and security assessments be done in an integral fashion and realistically model plant and preparedness response. The entire process ensures that safety and security design issues and preparedness requirements are

---

[2]Excerpt from the Atomic Energy Act: Sec. 3. Purpose.
It is the purpose of this Act to...[provide] for–
a.   a program of conducting, assisting, and fostering research and development in order to encourage maximum scientific and industrial progress;
b.   a program for the dissemination of unclassified scientific and technical information and for the control, dissemination, and declassification of Restricted Data, subject to appropriate safeguards, so as to encourage scientific and industrial progress;
c.   a program for Government control of the possession, use, and production of atomic energy and special nuclear material, whether owned by the Government or others, so directed as to make the maximum contribution to the common defense and security and the national welfare, and to provide continued assurance of the Government's ability to enter into and enforce agreements with nations or groups of nations for the control of special nuclear materials and atomic weapons.
d.   a program to encourage widespread participation in the development and utilization of atomic energy for peaceful purposes to the maximum extent consistent with the common defense and security and with the health and safety of the public;
e.   a program of international cooperation to promote the common defense and security and to make available to cooperating nations the benefits of peaceful applications of atomic energy as widely as expanding technology and considerations of the common defense and security will permit; and
f.   a program of administration which will be consistent with the foregoing policies and programs, with international arrangements, and with agreements for cooperation, which will enable the Congress to be currently informed so as to take further legislative action as may be appropriate.

addressed early in the design and that the interface among safety, security, and preparedness be considered in decisions on plant design, operations, and security.

The NRC's *safety expectations* are anchored in the Commission's safety goals [NRC 1986], which are based on the idea of minimizing additional risk burden to the population for the benefits of nuclear power. These underlying ideas are as appropriate for future reactors as they are for existing light water reactors (LWRs).

The Commission in their policy statement on "Regulation of Advanced Nuclear Power Plants" [NRC 1994], expected that:

•   Advanced reactors will provide enhanced margins of safety,

•   Advanced reactor designs will comply with the Commission's safety goal policy statement.

Accordingly, the framework is using the NRC safety goal Quantitative Health Objectives (QHOs) as the level of safety that the requirements are intended to achieve.

NRC's *security expectations* are that advanced reactors will provide enhanced margins of safety and utilize simplified, inherent, passive or other innovative means to accomplish their safety and security functions. To implement these expectations, the Framework proposes the following with respect to what security at nuclear power plants is to achieve. As such, the level of safety and security to be achieved, the scope of what needs to be protected and considered, and key aspects of the approach to be followed are expressed. They also provide guidance on the scope and purpose of the security performance standards to be developed.

Specifically, the security expectations for future plants encompass the following:

•   Protection of public health and safety, the environment and the common defense and security with high assurance is the goal of security.

•   The overall level of safety to be provided for security-related events should be consistent with the Commission's expectations for safety from non-security-related events.

•   Security is to be considered integral with (i.e., in conjunction with) safety and preparedness.

•   A defined set outside the DBT are to be considered, as well as the design basis threats (DBT), to identify vulnerabilities, and provide margin.

•   Defense-in-depth is to be provided against the DBT and each event outside the DBT considered, to help compensate for uncertainties.

•   Security is to be accomplished by design, as much as practical.

The NRC's *preparedness expectations* include the necessity for emergency preparedness capability, regardless of reactor technology or design or level of safety. On-site and off-site preparedness is expected to be able to support the response to the full range of accidents and security threats. The objective of emergency preparedness is to simplify decision making during emergencies. The emergency preparedness process incorporates the means to rapidly identify, evaluate, and react to a wide spectrum of emergency conditions. Actions, such as planning and coordination meetings, procedure writing, team training, emergency drills and exercises, and pre-positioning of emergency equipment, all are part of "emergency preparedness." Emergency plans are expected to be dynamic and routinely reviewed and updated to reflect an ever changing environment. The NRC expects that an acceptable, integrated emergency plan will be in place that provides reasonable assurance that adequate protective measures can, and will, be taken in a radiological emergency.

The requirements for emergency planning established in 10 CFR Part 50 and associated guidance will be applicable to new reactors unless specific changes can be justified. Making emergency preparedness more risk-informed and performance-based, may be a way to justify changes. A risk-informed and performance-based emergency preparedness regulatory structure could be more efficient and could free up resources. With a performance-based approach, licensees and communities would have the flexibility to address design specific challenges and develop their own unique solutions.

Chapter 3 discussed these high level goals and expectations in detail.

2.3   Element 2: Defense-in-Depth

The next element, the first step in developing the framework, is to address defense-in-depth. In the past, it has been an inherent part of NRC's regulatory structure; however, for future licensing, what should be the role of defense-in-depth? How should defense-in-depth be factored into the framework? What is its purpose? How is defense-in-depth related to uncertainties?

A core principle of the NRC's safety philosophy has always been the concept of defense-in-depth, and defense-in-depth remains basic to the safety, security, and preparedness expectations in the framework. "The defense-in-depth philosophy ......has been and continues to be an effective way to account for uncertainties in equipment and human performance." The ultimate purpose of defense-in-depth is to compensate for uncertainty (e.g., uncertainty due to lack of operational experience with new technologies and new design features, uncertainty in the type and magnitude of challenges to safety).

In licensing future reactors, the treatment of uncertainties will play a key role in ensuring that safety limits are met and that the design is robust for unanticipated factors. Defense-in-depth is now integrated into the Framework in that the subsequent elements are developed based on the criteria established for defense-in-depth, as shown in Figure 2-3.

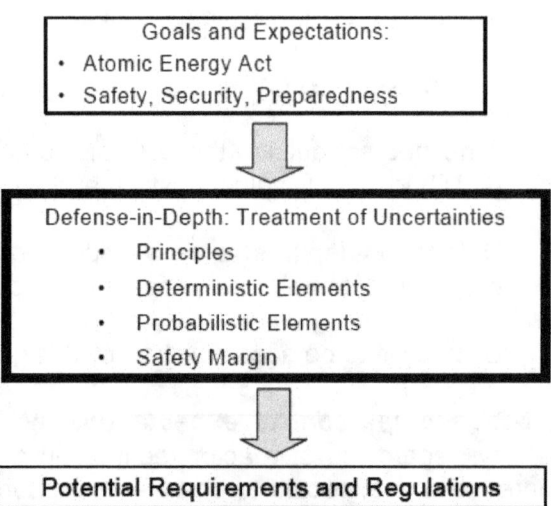

Figure 2-3   Integration of Defense-in-Depth in Development of Framework

The aim of the framework is to develop an approach to defense-in-depth for future reactors that is consistent with the successful past practices used for operating reactors, but which improves on past practices by being more consistent and by making use of quantitative information where possible.

The deterministic elements of the framework include protective strategies (see Section 2.4 below) and a set of defense-in-depth principles are followed. These principles[3] are established by examining the different kinds of uncertainties to be treated, and incorporating successful past practices and lessons learned related to defense-in-depth. They are applied regardless of the likelihood of failure. Therefore, characteristics of this approach are that

- The high level lines of defense are maintained (protective strategies),
- Accident prevention alone cannot be relied on to reach an acceptable level of safety, and
- The capability to mitigate accidents is also needed.

This approach to defense-in-depth is used primarily to address completeness uncertainties.

The probabilistic elements of defense-in-depth are the aggregate of provisions made to compensate for uncertainty and incompleteness in our knowledge of accident initiation and progression. The probabilistic element acknowledges probabilistic risk assessment (PRA) as a powerful tool in the search for the unexpected and the identification of uncertainties.

The probabilistic elements also establish adequate defense-in-depth measures, including safety margins, to compensate for those scenarios and their uncertainties which are quantified in the PRA model. The ability to quantify risk and estimate uncertainty using PRA techniques and taking credit for defense-in-depth measures in risk analyses, allows a better answer to the question of how much defense-in-depth is enough, given the likelihood of the event and its uncertainty.

Safety margin is as an integral part of defense-in-depth, since the basic purpose of safety margins is to cope with uncertainty. For example, consensus codes and standards contribute to defense-in-depth because they incorporate safety margins. In addition, the compensatory measures that are also part of defense-in-depth need to have some margins embedded in them to deal with the accident and malfunction. Compensatory measures that lack margin are not very practical.

Chapter 4 discusses the approach to defense-in-depth, the principles and criteria, in detail.

2.4 Element 3: Safety Fundamentals

This element provides the path, or process, from the high level goals and expectations to actually establishing specific requirements. For example, what overall process should be used to identify potential requirements for design, construction, and operation? Should the process to identify the specific requirements and regulations start with design considerations? Or, should the process start with some other consideration?

---

[3]The Defense-in-Depth Principles discussed in Section 4.3 are: (1) Measures against intentional as well as inadvertent events are provided, (2) The design provides accident prevention and mitigation capability, (3) Accomplishment of key safety functions is not dependent upon a single element of design, construction, maintenance, or operation, (4) Uncertainties in SCCs and human performance are accounted for in the safety analyses, (5) The design has the capability to prevent an unacceptable release of radioactive material, (6) Plants are sited at locations that facilitate the protection of public health and safety.

The process chosen to initially identify and define the requirements and regulations needs to implement the safety, security, and preparedness expectations and ensure protection of the public health and safety. Safety fundamentals have been defined, using a defense-in-depth approach, in the form of protective strategies that, if met, will ensure the protection of the public health and safety with a high degree of confidence. That is, these protective strategies are directly tied to meeting the high level goals and expectations. The framework identifies five protective strategies: Physical Protection, Stable Operation, Protective Systems, Barrier Integrity, and Protective Actions as put in context in Figure 2-4. The protective strategies introduced here set the design, construction, and operating conditions that will ensure protection of public health and safety, workers, and the environment.

Figure 2-4   Framework Protective Strategies

Defense-in-depth is integrated into the approach defining the Protective Strategies. The approach is based on a philosophy of regulation that requires multiple strategies (i.e., lines of defense) to ensure that there is little chance of endangering public health and safety. It is a top-down, hierarchical approach that starts with a desired outcome and identifies protective strategies (safety fundamentals) to ensure this outcome is achieved even if some strategies should fail. The protective strategies provide defense-in-depth that offer multiple layers of protection of public health and safety.

- The **Physical Protection** objective is to protect workers and the public against intentional acts (e.g., attack, sabotage, and theft) that could compromise the safety of the plant or lead to radiological release.

- The **Stable Operation** objective is to limit the frequency of events that can upset plant stability and challenge safety functions, during all plant operating states, i.e., full power, shutdown, and transitional states.

- The **Protective Systems** objective is to ensure that the systems that mitigate[4] initiating events are adequately designed, and perform adequately, in terms of reliability and capability, to satisfy the design assumptions on accident prevention and mitigation during all states of reactor operation. Human actions to assist these systems and protect the barriers are included here.

---

[4]Protective systems provide a mitigation role by features and capabilities that fulfill safety functions in response to initiating events and thereby protect the barriers. They also provide a prevention role by application of design and operational features that contribute to their reliability and thereby reduce the probability that an initiating event will lead to an accident involving protective systems failures.

- The **Barrier Integrity**[5] objective is to ensure that there are adequate barriers to protect the public from accidental radionuclide releases from all sources. Adequate functional barriers need to be maintained to protect the public and workers from radiation associated with normal operation and shutdown modes and to limit the consequences of reactor accidents if they do occur. Barriers can include physical barriers as well as the physical and chemical form of the material that can inhibit its transport if physical barriers are breeched.

- The **Protective Actions** objective is to ensure that adequate protection of the public health and safety in a radiological emergency can be achieved should radionuclides penetrate the barriers designed to contain them. Measures include emergency procedures, accident management, and emergency preparedness.

A top-down analysis of each protective strategy leads directly to a categorization of the kinds of requirements that can ensure that the protective strategies are met. The protective strategies provide the primary basis for the development of the requirements, as introduced in Section 2.6.

Chapter 5 describes the manner in which these protective strategies are met.

2.5   Element 4: Licensing Basis

A major goal is that the regulatory licensing basis be risk-informed. The current regulatory structure is deterministic and is being modified in places to incorporate risk insights. A risk-informed regulatory structure should integrate risk from conception. For example, what should be the balance between deterministic and probabilistic elements of the licensing process? Should a deterministic set of requirements first be defined and then refined with risk insights, or should a set be defined using deterministic and probabilistic criteria in an integrated fashion?

In the current Part 50, the licensing basis is established with a deterministic approach and an LWR focus. Consequently, a stylized set of accidents to be considered are not necessarily risk or design applicable. In the Framework, probabilistic criteria integrated with deterministic criteria based on plant-specific considerations are used to establish the potential requirements. Figure 2-5 shows this new element to the framework.

In using a probabilistic process to establish the licensing basis, confidence in the technical acceptability becomes a key factor. Therefore, the technical acceptability of the PRA is part of this element, and this document addresses this issue.

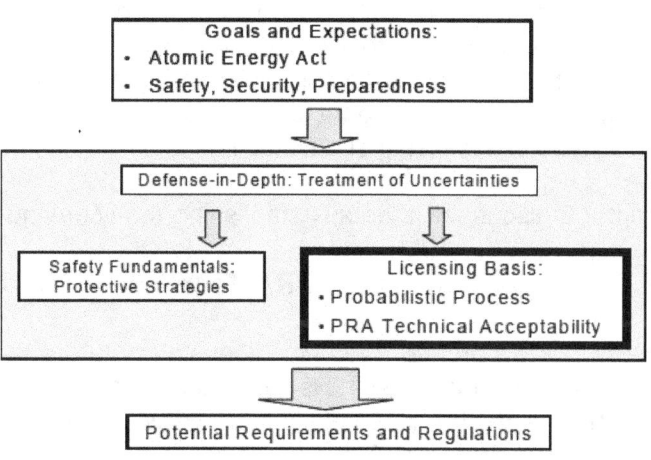

Figure 2-5   Framework Licensing Basis

---

[5]Note that the purpose of barriers, protective systems and emergency preparedness is to mitigate the accident sequences by reducing their frequency or their impact. Historically engineers have spoken of preventing core melt and mitigating core damage. These terms are not especially helpful with some future reactor designs and prevention/mitigation definitions change as the object under discussion changes - core damage, release from the primary system, release off-site.

2.5.1 Licensing Basis: Probabilistic Process

The licensing basis criteria are parallel and complementary with the Protective Strategies, in support of the NRC's defense-in-depth expectations, as shown in Figure 2-5. The probabilistic criteria include compliance with the quantitative health objectives (QHOs) of the NRC's safety goals.

The framework establishes probabilistic criteria to ensure that

- The integrated plant risk is acceptable in terms of the QHOs of the NRC's safety goal policy statement,

- A frequency consequence (F-C) curve is developed (Chapter 6) that together with the plant PRA is used to select licensing basis events.

- The selection of those events that are used to establish the licensing basis of the design (licensing basis events or LBEs) is carried out in a risk-informed manner,

- The LBEs meet the F-C curve with margin, and

- The safety classification of systems, structures, and components (SSCs) reflects their importance in reducing plant risk.

In selecting both the LBEs and the safety significant SSCs, defense-in-depth measures are incorporated, but, in addition, the risk information from the PRA is used to focus attention on the risk-significant aspects of the design.

LBEs derived from the PRA need to meet stringent probabilistic acceptance criteria and, depending on their frequency, need to meet additional deterministic (defense-in-depth) criteria. In this manner, the LBEs provide additional assurance that the design has adequate defense-in-depth in the form of sufficient margin to account for uncertainties. The LBEs also include a deterministically selected event, used in assessing site suitability.

Chapter 6 discusses in detail establishing the licensing basis using probabilistic criteria.

2.5.2 Licensing Basis: PRA Technical Acceptability

PRA will play a greater role in the licensing of future reactors. Since the quality of the PRA used in making licensing decisions is commensurate with the significance of the regulatory decision, the expectations of the quality of future PRAs is greater than for current operating plants. This document requires a PRA during the pre-application, design certification, one-step (i.e., combined operating license) and two-step (i.e., construction permit and operating license) license reviews, and during plant operation. The PRA is used in this NUREG to help in the following:

- Establishing the LBEs,

- Identifying safety-significant SSCs and their corresponding special treatment requirements,

- Identifying key uncertainties and associated research needs to address them,

- Developing plant operating procedures, technical specifications, configuration control procedures, and emergency response and accident management plans,

- developing inspection, surveillance, maintenance, and monitoring programs

- Comparing the plant design and operation against licensing risk criteria.

This document identifies the requirements necessary to ensure the quality of PRA necessary for the use of the PRA in licensing applications. High-level requirements (HLR) are provided for evaluating both internal and external events during all modes of operation. In meeting these HLRs, many of the current PRA methods, techniques, and data used for LWRs apply. However, modeling future reactors may necessitate extensions of current PRA methods.

For example, future reactor designs may focus on the use of passive systems and inherent physical characteristics to ensure safety, rather than relying on the performance of active electrical and mechanical systems. For plants, with many passive systems, fault trees may be very simple when events proceed as expected and event sequences may appear to have very low frequency. The real work of PRA for these designs may lie in searching for unexpected scenarios. Innovative ways to structure the search for unexpected conditions that can challenge design assumptions and passive system performance will need to be developed or identified and applied to these facilities. The risk may arise from unexpected ways the facility can end up operating outside the design assumptions. For example, there may be a need for a hazard and operability analysis (HAZOP) style search scheme for scenarios that deviate from designers' expectations and structured search processes for construction errors, operator and maintenance errors, aging problems, and gradual degradation of passive systems.

This document builds on existing PRA quality requirements delineated in Regulatory Guide 1.200 [NRC 2007] and the PRA standards[6]. This document provides methods to help ensure PRA quality, including establishing PRA consensus standards that provide supporting requirements to the proposed high-level requirements, an independent peer review process, Regulatory Guides and Standard Review Plan guidance to assess PRA quality, and guidance on how to perform specific aspects of an advanced reactor PRA. The use of PRAs in the licensing and operation of future reactors will require that PRAs be living documents.

The technical acceptability of the PRA to support the risk-informed aspects of the Framework is discussed in detail in Chapter 7.

2.6   Element 5: Integrated Process

Various elements (in the form of goals, expectations, principles, criteria) have been defined. The relationship of these elements in actually defining a complete set of requirements for design, construction, and operation, is needed. For example, how should the elements in the framework be integrated? Should each element be treated and addressed separately and independently? Or, should they be integrated to achieve a cohesive set of requirements and regulations?

---

[6]ASME/ANS PRA Standard, "Standards for Level 1/Large Early Release Frequency Probabilistic Risk Assessment for Nuclear Power Plant Applications," covering internal and external events at full power for light water reactors to be issued in early 2008.

The process for identifying the requirements begins with the protective strategies. Each one is examined with respect to what are the various threats or challenges that could cause the strategy to fail. These challenges and threats are identified using a logic tree to perform a "systems analysis" of the strategy to identify potential failures. The defense-in-depth principles are then applied to each protective strategy. Defense-in-depth measures are identified which should be incorporated into the requirements to help prevent protective strategy failure. This approach forms the process for the selection of "topics." Requirements are then identified for each topic.

Part of the process involves development of guidance to be used for actually writing the requirements. This guidance addresses writing the requirements in a performance-based fashion, incorporating lessons learned from past experience, and utilizing existing requirements and guidance, where practical. The guidance also ensures that the probabilistic process for establishing the licensing basis are incorporated. All of the above are integrated and results in a set of potential requirements which serve to illustrate and establish the feasibility of developing a risk-informed and performance-based licensing approach.

This last "element" of the framework is shown in Figure 2-6. This integrated process is discussed in detail in Chapter 8.

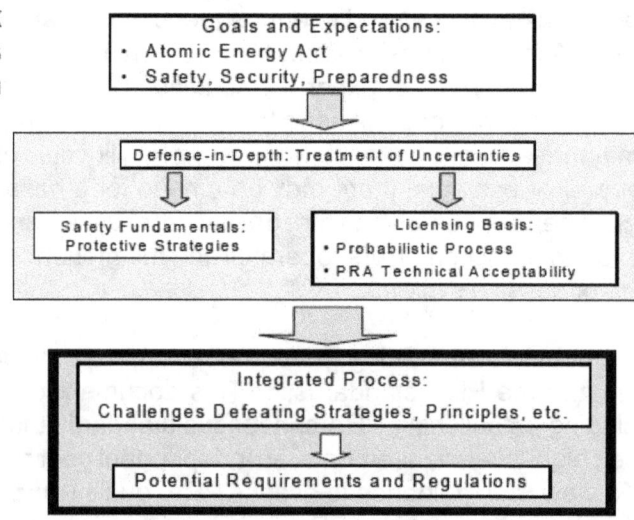

Figure 2-6   Framework Integrated Process

2.7   Summary

This chapter has provided an introduction to the complete Framework. Each of the elements of the Framework are discussed in detail in the subsequent chapters and ultimately come together and result in a set of potential requirements (Appendix J) that illustrates the feasibility of using the Framework to establish a risk-informed and performance-based set of requirements for future plant licensing.

2.8 References

[NRC 1986]   U.S. Nuclear Regulatory Commission, "Safety Goals for the Operations of Nuclear Power Plants: Policy Statement," 51 FR 30028, Washington, DC, August 4, 1986.

[NRC 1994]   U.S. Nuclear Regulatory Commission, Commission Policy Statement on the Regulation of Advanced Nuclear Power Plants, 59 FR 35461, July 1994.

[NRC 2007]   U.S. Nuclear Regulatory Commission, "An Approach for Determining the Technical Adequacy of Probabilistic Risk Assessment Results for Risk-Informed Activities," Regulatory Guide 1.200, Revision 1, January 2007.

[US 1954]   Public Law 83-703, The Atomic Energy Act of 1954, as Amended, August 1954.

3. GOALS AND EXPECTATIONS: SAFETY, SECURITY, AND PREPAREDNESS

3.1 Introduction

The purpose of this chapter is to define the goals and expectations for safety, security, and preparedness required at future plants licensed by the U.S. Nuclear Regulatory Commission (NRC). The chapter stresses the integrated nature of all three: any change to one of them requires consideration of the impact on all. The framework was developed following a top-down approach, as shown in Figure 3-1. It is built upon the traditional NRC safety mission, beginning with the Atomic Energy Act and encompassing a set of safety, security, and preparedness goals expectations developed from Commission statements.

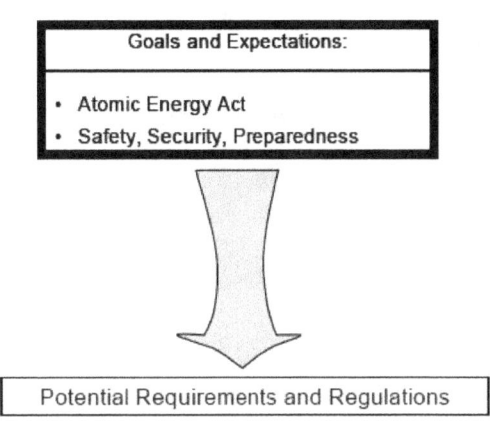

Given that the basic approach in developing the framework is to start with a "clean sheet of paper," the next question would be where to start the structure. For example, should the structure be based on a high-level, top-down approach starting with some basic principles and goals? Or, should the framework be more of a bottom-up approach starting with fundamental design criteria?

Figure 3-1   Framework Approach

An optimum approach is an hierarchal top-down structure. It would start at the highest level in defining the basic goals and criteria: the Atomic Energy Act, which provides the mission to ensure that commercial nuclear power plants are operated in a manner that provides adequate protection of public health and safety and is consistent with the common defense and security. The other top level criteria address the Commission's expectations for safety, security and preparedness.

This NUREG integrates the expectations for safety, security, and preparedness to achieve the overall level of safety expected by the NRC. The approach is based on safety and security assessments that realistically model plant and preparedness response; the results of these assessments become part of regulatory requirements. The entire process ensures that safety and security design issues are addressed early in the design development and regulatory review process, so that the resulting design relies more on inherent design characteristics and features and less on extrinsic operational safety and security programs. The Framework is designed to ensure meeting the NRC's safety and security expectations on a technology-neutral basis, i.e., a licensing basis that can be applied to all new plants, regardless of technology.

The NRC's *safety expectations* are anchored in the safety goals, which seek to minimize additional risk burden to the population for the benefits of nuclear power. These underlying ideas are as appropriate for future reactors (or any new technology) as they are for existing light water reactors (LWRs). The NRC's *security expectations* are discussed in SECY-07-0167, "Revision of Policy Statement on Regulation of Advanced Reactors," [NRC 2007]. The approach anticipates that establishment of security design aspects early in the design process will result in a more robust and effective security posture and less reliance on human actions. New reactors need to be protected at least as well as currently operating plants [NRC 2002, NRC 2003]. The NRC's *preparedness expectations* include the necessity for an emergency planning and preparedness capability, regardless of reactor technology or design.

3.2 Safety Expectations

3.2.1 Level of Safety

The level of safety that future reactors are expected to meet are the risk objectives, i.e., the quantitative health objectives (QHOs), embedded in the NRC's safety goal policy.

The Commission in their Policy Statement on "Regulation of Advanced Nuclear Power Plants," expects that:

(1)   Advanced reactors will provide enhanced margins of safety,
(2)   Advanced reactor designs will comply with the Commission's safety goal policy statement.

To show the relation of the safety goal risk objectives to future plant licensing, and to address the Commission's expectations, a three-region approach to risk tolerability/acceptance is developed and defined as illustrated in Figure 3-2.

High risk

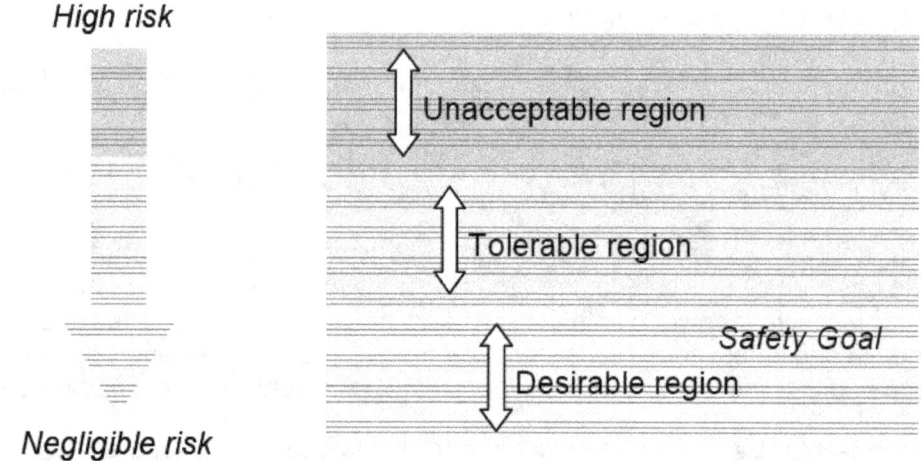

Negligible risk

Figure 3-2   Three Region Approach to Risk Tolerability/Acceptance

Such a Three-region approach to risk acceptance for nuclear power plants, including operating reactors, has been discussed and used in several forums. In considering this figure, the substantial uncertainty (see Chapter 4 for a discussion of uncertainty) in a plant's risk performance is taken into account. Conceptually, the lowest region represents the value of the risk metric that corresponds to the desired safety goal and/or risk objective. This region defines what is "safe enough", i.e., a region in which further safety enhancements are not needed.[7]

The objective of this document is to help develop requirements for future reactors that are consistent with the risk lying in the lower, desirable, risk region, i.e., requirements that will ensure there is only a small chance that the risk will extend into the tolerable region, and a negligible chance that it reaches the upper, unacceptable region. Accordingly, *the regulatory requirements under the Framework for new reactors are established to keep the risk in the desirable*

---

[7]Figure 3-2 is conceptual. The detailed considerations that would be necessary to implement this idea on a quantitative basis are discussed in Chapter 6.

region, that is, the regulations are written to achieve the safety goal level of safety. The rationale for these requirements is twofold: they provide enhanced margins to account for uncertainties, in particular, those that may be associated with new designs and technologies, and they help implement the Commission's expectations for enhanced safety as expressed in the Advanced Reactor Policy Statement.

Adoption of the QHOs as the basis for the level of safety implies an increase in safety for future reactors compared to current LWR designs. However, the QHOs have been used to assess current LWR vulnerabilities and safety improvements have been made, where justified. Accordingly, current reactors are more safe than required by the letter of the regulations and, in many cases, achieve a level of safety comparable to the QHOs. Therefore, developing requirements that make the level of safety correspond to the QHOs ensures that future plants "comply with the Safety Goal Policy Statement," as stated in the Commission's Policy Statement on "Regulation of Advanced Nuclear Power Plants" and imparts stability and consistency to the regulatory process. Finally, the approach of this document is consistent with industry initiatives, which are directed at developing designs with enhanced safety over currently operating plants.

The above discussion should not be taken to imply that comparing a plant's safety profile to the QHOs can be accomplished with little uncertainty. The current probabilistic risk assessment (PRA) technology is relatively mature for estimating the risk from internal events for LWRs operating at full power. Techniques for estimating the risk related to fire, external events, and other modes of operation for these types of reactors are less mature. Furthermore, for non-LWRs the state-of-the-art in PRA is less advanced than for LWRs. Finally, estimating the risk from deliberate adversarial acts of theft, sabotage, and/or attack is very difficult. All these factors argue for the need to compensate for the significant uncertainties encountered in comparing the plant safety profile to the QHOs via the 'margins' implied in Figure 3-2 between adequate protection and the safety goals, and by the application of defense-in-depth as discussed in Chapter 4 of this report.

3.2.2 Implementing the NRC's Safety Expectations

With the level of safety expressed in terms of risk, PRA plays a significant role in the framework. With the framework approach, the PRA information is used during the design stage and for selecting licensing basis events (LBEs) and safety significant systems, structures and components (SSCs). Therefore, the scope of the PRA encompasses the whole spectrum of events that can credibly occur during the life of the plant: normal operation, as well as frequent, infrequent, and rare initiating events and accident event sequences. This scope is broader than that used currently for LWR risk analysis, which concentrates on beyond design basis accidents, i.e., accidents leading to severe core damage.

For current LWRs, the risk examined is almost always expressed in terms of the surrogate risk objectives: core damage frequency (CDF) and large early release frequency (LERF). One reason why CDF and LERF provide adequate risk metrics for the current LWRs is that it can be demonstrated[8] that these parameters, namely, CDF and LERF, can be used as surrogate metrics for the NRC's safety goal QHOs. For new, non-LWR types of reactors, however, not only will the quantitative values for CDF and LERF be no longer applicable, CDF itself may no longer be a useful risk metric.

---

[8]CDF and LERF were found to be surrogates for the QHOs for LWRs based on the amount and characteristics of the radionuclide inventory in currently operating LWRs, the timing and magnitude of potential releases in severe accidents at these plants, and the anticipated and planned emergency response of the nearby offsite population to these releases at operating sites.

Risk metrics applicable to a variety of different reactor designs, are either ones that express consequences directly or that can be linked to consequences without technology-specific metrics. Dose to the public from radiological releases is an example of a metric that is closely linked to consequences, and therefore is one of the parameters used to express PRA results. In addition,

since frequent, infrequent, as well as rare events are included in the PRA, a single limiting criterion (such as CDF for LWRs) is not adequate. Instead, a criterion that specifies limiting frequencies for a spectrum of consequences, from none to very severe, needs to be established. This can be denoted via a frequency consequence (F-C) curve, (which provides guidance to a plant designer). Figure 3-3 shows an example.

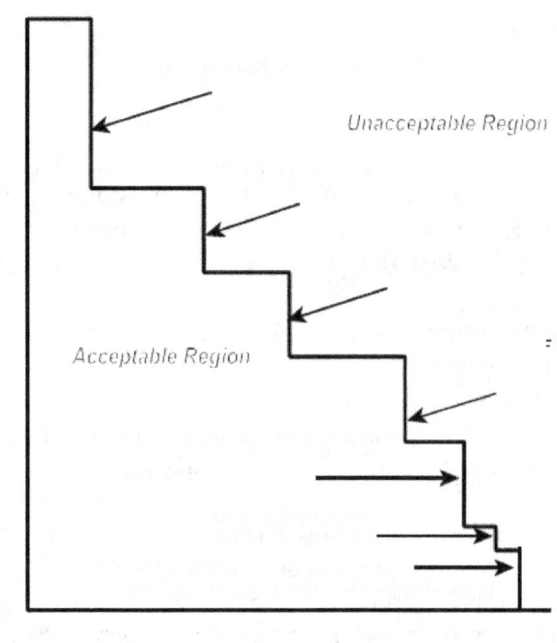

As explained in more detail in Chapter 6, this F-C curve relates the frequency of potential accidents to acceptable radiation doses at the site boundary from these accidents. It is based on, and derived from, current regulatory requirements in Parts 20, 50, and 100 of Title 10 of the Code of Federal Regulations (10 CFR). 10 CFR Part 20 limits the radiation doses from licensed operation to individual members of the public. 10 CFR Part 50 Appendix I identifies design objectives for releases during normal operation to be as low as

Figure 3-3   Frequency-Consequence Curve

reasonably achievable (ALARA). Part 50.34 requires an applicant for a license for a power reactor to demonstrate that doses at the site boundary (and the outer boundary of the low population zone) from hypothetical accidents will meet specified criteria. Likewise, 10 CFR Part 100 has similar dose criteria for determining site suitability. The principle underlying the F-C curve is that event frequency and dose are inversely related, i.e., the higher the dose consequences, the lower is the allowed event frequency.

The sequences of the PRA populate the space under the F-C curve. Some scenarios will have little or no consequences, primarily because of the inherent characteristics and design features of the plant. Others are likely to approach the F-C curve and thus make up the important contributors to the plant risk profile. To be acceptable, the results of the PRA, in terms of the frequency and consequences of all the accident sequences examined, need to lie in the acceptable region, (i.e., below) the F-C curve. The accident sequences will be used to select the licensing basis events (LBEs).

The risk in the safety goals and the QHOs is the total plant risk incurred over a reactor year. This means the PRA results need to demonstrate that the total plant risk, i.e., the risk summed over all of the accident sequences in the PRA, need to satisfy both the latent cancer QHO and the early fatality QHO. The safety goals, and consequently, the QHOs are phrased in terms of the risk to an 'average' individual in the vicinity of (or 'area near') a nuclear power plant per reactor year. The latent cancer QHO is defined in terms of the risk to an average individual within 10 miles, and the early fatality QHO in terms of the risk to an average individual within 1 mile of the plant.

Note that with the kind of acceptance criterion for individual sequences described above, an accident sequence is acceptable even though it has a dose at the boundary associated with it, as long as its frequency does not exceed the limit for that dose, as specified by the F-C curve. For the PRAs required here, whose scope covers all types of off-normal event sequences, the criterion is a series of limiting frequencies whose value depends on the associated consequences; frequent event sequences need to lead to no consequences or very minor ones; infrequent event sequences can have somewhat higher doses associated with them, and rare events can have higher consequences still. For specific technologies it may be possible to eventually develop surrogate metrics (such as CDF and LERF for LWRs) for the dose parameter, along with acceptable values for such surrogates.

In summary, in the framework approach to licensing, a PRA is used to generate a sufficiently complete set of LBEs, and to provide an estimate of the overall risk profile of the plant. The framework advocates a PRA as the best available analysis method for showing how the interactions and dependencies among SSCs, human actions, and potential plant hazards can result in accident sequences being initiated, prevented, and mitigated. The scope of the PRA will differ from the current LWR PRAs. Uncertainties need to be addressed in the calculation of both frequencies and consequences of the accident scenarios. Since the accidents include rare events postulated to occur in complex systems for which limited experience exists, the consideration of uncertainties are a vital part of understanding the extent of the risk (and of selecting the LBEs). The PRA information, together with a suitable F-C curve, is used to select LBEs and safety significant SSCs.

3.2.3   Surrogate Risk Objectives

The Commission's overall expectation for protection of public health and safety from accidents resulting from nuclear power plant operation is expressed in its 1986 Safety Goal Policy Statement. The goal of the framework for new plant licensing is to ensure that new plants achieve a level of safety at least equivalent to that expressed by the Safety Goal Policy Statement. For currently operating LWRs, surrogate objectives related to core damage prevention and accident mitigation, (i.e., core damage frequency (CDF) and large early release frequency (LERF) or conditional containment failure probability (CCFP), have been developed and used as surrogates for the quantitative health objectives (QHOs) expressed in the Safety Goal Policy Statement (i.e., 2×10^{-6}/ry individual risk for latent fatalities and 5×10^{-7}/ry individual risk for early fatalities). These surrogate objectives focus on plant design and have eliminated the need for carrying out probabilistic consequence analysis in PRAs for currently operating LWRs.

These LWR specific surrogate risk objectives have been used as the basis for various risk-informed activities for currently operating plants. The numerical values used for these surrogates (10^{-4}/ry for CDF, 10^{-5}/ry for LERF, and 0.1 for CCFP) are based upon the characteristics and risk analysis associated with currently operating light-water reactor plants (e.g., plant size, performance, source term, emergency preparedness) and their site characteristics (i.e., meteorology and population distribution). In effect, for current LWRs the 10^{-4}/ry CDF serves as a surrogate for the latent fatality QHO as well as a measure of accident prevention, and the 10^{-5}/ry LERF or 0.1 CCFP serves as a surrogate for the early fatality QHO for currently operating reactors. (See Appendix D for detailed discussion on derivation of surrogates for currently operating LWRs).

As discussed earlier, a frequency-consequence curve was established to support achievement of the overall safety objective of the technology-neutral licensing approach and to define acceptance criteria for individual accident sequences in the PRA and for those accident sequences chosen as LBEs. This frequency-consequence curve is anchored in the safety goal QHOs and other existing requirements and is defined in terms of dose to an individual at specific distances that are defines in Chapter 6, e.g., the site boundary and the low population zone. Accordingly, a probabilistic consequence analysis is needed to implement the F-C curve. However, this frequency consequence curve is not a substitute for the QHOs, which express goals for the cumulative latent and early fatality risk from accidents and also require a level 3 PRA analysis. Given the frequency-consequence curve and the QHOs, it is useful to ask: (1) if technology neutral surrogate risk objectives would be useful as substitutes for the QHOs; (2) if so, how would they be used; and (3) what should they be? Each is discussed below.

Although the frequency-consequence curve is anchored in the safety goal QHOs, it is not itself a measure of compliance with the QHOs (i.e., as mentioned above, the frequency-consequence curve is not an assessment of the cumulative risk from all event sequences considered in the design). Accordingly, to ensure the QHOs are met, either a probabilistic consequence assessment that calculates offsite early fatality and latent cancer risks is needed or, if possible, technology-neutral surrogate risk objectives are developed that can account for the cumulative risk from all event sequences considered in the design and reduce the need for a probabilistic offsite consequence assessment . To be most useful, these surrogate objectives should also focus more directly on plant design, thus simplifying the analysis needed.

Surrogate risk objectives, if they can be defined in a meaningful technology-neutral way, can also be useful in defining the desired apportionment between accident prevention and accident mitigation (which is not defined by the QHOs or the frequency-consequence curve), as is done today for currently operating LWRs, via the use of CDF and LERF. Defining such a apportionment quantitatively will be useful in implementing the defense-in-depth principles discussed in Chapter 4.

It is envisioned that technology-neutral surrogate risk objectives would be used in the following ways:

(1)   as a simplified way to assess the design's compliance with the QHOs;

(2)   as quantitative measures to implement the defense-in-depth principle on accident prevention versus mitigation;

(3)   as the top level criteria for establishing reliability goals for protective systems and consistent with initiating event frequency;

(4)   as a probabilistic counterpart to the deterministic criteria directed toward accident prevention (discussed in Chapter 6).

CDF and LERF can be demonstrated to be acceptable surrogate risk objectives for current LWRs. Such a demonstration depends on the characteristics of LWR technology, in particular, the ways in which severe accidents can occur and the source terms related to these accidents. Given these characteristics, one can show that restricting CDF below 10^{-4}/ry and LERF below 10^{-5}/ry will ensure that the consequence limits embodied in the latent cancer and early fatality safety goals can be met. Surrogate objectives for future reactors will, in general, be technology-dependent, and it is unlikely that surrogate measures similar to CDF and LERF could be identified on a technology-neutral basis. Even something as simple as defining an accident in terms of radionuclide release can be difficult on a technology neutral basis. See Appendix C for a more thorough discussion.

3.3   Security Expectations

Security expectations for advanced reactors are stated in SECY-07-0167. The framework proposes a major difference in approach for new plants, a risk-informed approach, where security will be evaluated integral with the design, rather than post-design compensatory measures. As discussed in Section 6.7, the security expectations for new plants encompass the following:

•   Protection of public health and safety, the environment, and common defense and security with high assurance is the goal of security.

•   The overall level of safety to be provided for security-related events should be consistent with the Commission's expectations for safety from non-security related events.

•   Security is to be considered integral with (i.e., in conjunction with) design and preparedness.

•   A defined set of events outside of the design basis threats (DBTs) should be considered, as well as the DBT, to identify vulnerabilities and provide margin.

•   Defense-in-depth should be provided against the DBT and each event outside the DBT considered, to help compensate for uncertainties.

•   Security is to be accomplished by design, as much as practical.

The above security expectations define the elements addressed in the security performance standards (described in Section 6.7). These security expectations are intended to promote enhanced security, emphasize design solutions to security issues, provide means to ensure integration of security, safety and preparedness and provide guidance for qualitative and quantitative measures for assessing security. The amount of intrinsic protection and the ways in which it can be built into the design will depend on technology-specific features of the design.

A security assessment will be required to ensure performance is adequate. The security assessment will involve characterization of the potential threat, the potential targets, and the potential consequences. The frequency of the threat depends on some factors only known to the adversary; therefore conditional risk will play a key role in the proposed security performance standards.

3.4   Preparedness Expectations

The NRC's basic preparedness expectation is to ensure the capability to effectively protect the public through dose-saving, should substantial amounts of radioactive material be released to the environment.

Criteria for determining the scope of required off-site emergency preparedness measures are needed that address technology-specific factors, such as reactor size (power level), location, level of safety (i.e., likelihood of release), magnitude and chemical form of the radionuclide release, and timing of releases. Some conditions and considerations affecting the required response that have particular importance would include the following:

3. Goals and Expectations

(1)   consideration of the full range of accidents
(2)   use of defense-in-depth
(3)   prototype operating experience
(4)   acceptance by federal, state, and local agencies
(5)   acceptance by the public
(6)   security considerations

In a more general and integrated sense, preparedness takes many forms and the NRC's preparedness expectations include preparedness for safety and security.   Both kinds of preparedness involve both on-site and off-site activities, as shown in Figure 3-4.

Safety Preparedness on-site involves efforts to deal with conditions in the plant and communication/coordination with off-site entities by both operators and plant management .   It requires that operators are trained on the use of emergency and abnormal operating procedures and that these procedures are validated for the wide range of conditions under which they might be applied.   Likewise senior supervisory personnel need to understand the more conceptual accident management guidelines, exercise interactions with vendors, regulators, and other officials. Safety preparedness off-site involving traditional Emergency Preparedness fits within this structure.

Integrated Preparedness

•   Safety preparedness
      –   On-site measures
            ■   main control room [emergency operating and abnormal procedures]
            ■   emergency response center [accident management guidelines]
      –   Off-site emergency preparedness

Figure 3-4   Generalized Preparedness as Part of the Protective Actions Protective Strategy

Security Preparedness also has on-site and off-site elements.   On-site activities include the preparedness of the guard force, maintenance of design features that enhance physical protection, and the operators readiness to isolate damaged equipment.   Off-site security preparedness involves traditional Emergency Preparedness and also coordination with local and national police and first responder personnel.

This approach acknowledges that we need preparedness for both safety and security, both on-site and off-site.   It acknowledges that preparedness involves procedures and training, as well as hardware and software.

A key feature in preparedness is applying a graded approach in which response plans and procedures are tailored to the hazard or threat they are meant to neutralize or effectively respond to hazard or threat.

3.5   Integration of Safety, Security, and Preparedness

Safety, security, and preparedness are expected to function in a unified and coordinated manner. The expectations for the level of safety and protection, implemented through an integrated process, lead to a plant that is safe and robust against all internal and external hazards, against inadvertent and advertent threats, and meet the NRC objectives of protecting the public health and safety, and the common defense and security.

The plant's safety and security features are to be designed to minimize adverse interactions and optimize their integrated benefit. This includes potential SSC interactions that are the result of the design process, and expected operational and maintenance practices that could change the required response by operation or security personnel. An example of a design interaction is restricting the number of access points for a given location which may be beneficial to security, but could result in longer response times for operators during plant emergencies. An example of a operational interaction could be the removal from service of a key plant component for maintenance that could impact the plant's security response. Therefore, during the development of the safety and security requirements their impact on each other and on preparedness be examined. Changes in design, operation and maintenance that have the potential for adverse effects on safety and security, including the site emergency plan, will be assessed and managed before implementing changes to plant configurations, facility conditions, or security. This assessment could be accomplished using PRA methods and security vulnerability assessment techniques to provide a rational bases for decision-making by infusing safety objectives with security concerns.

The results of the safety and security assessment should be reviewed for the security performance standards and other considerations important to making a decision on the adequacy of security. This would require an integrated decision-making process that takes into account several factors that could influence the decision. This means that the plant should meet guidelines for both early and latent fatality risk, should meet defense-in-depth provisions of this NUREG, theft or diversion should be prevented by implementing the requirements in 10 CFR 73 and the post 9/11 orders, and design solutions to security-related issues should be developed, where possible, that eliminate the need for operational controls (e.g., operational security actions). Furthermore, the scope and quality of the analysis used in the security assessment should be consistent with the scope of the threat being assessed and with accepted methods and data, and the impact of security-related actions (e.g., design changes, operational changes) should not detract from overall plant safety or preparedness or worker safety. Unquantified uncertainties should be considered for whether they could have a major influence on the decision.

The decision process used in evaluating whether to make a change in plant design, operation or security as a result of the safety and security assessment would need to consider all of the above factors. Considering all of these factors will help to ensure integration of safety, security and preparedness. Ideally, all factors should be met before deciding to take an action. However, this may not always be possible, in which case the factors for and against a change should be weighed and the decision justified on a relative basis. As will be described in Chapter 4, this process requires the use of both probabilistic (risk) information and deterministic analysis. As discussed throughout this document, and especially in Chapter 4, defense-in-depth is the primary means for protecting against unknown factors affecting risk.

In addition to the integration of safety and security, the on-site and off-site preparedness is expected to be able to support the response to the full range of accidents and security threats. This results in the security and preparedness implications of design decisions being fully integrated with more traditional safety decisions. This integration needs to be exercised in such a way to ensure that security requirements do not place undue limits on the ability for preparedness in protecting plant safety, and the health and safety of nearby populations.

Another important element associated with integration is the expectation that there will be an increased reliance on design enhancements over operational solutions. These design enhancements should increase the margins associated with key elements of emergency response, including the required times and complexities of actions associated with operation, security, and emergency responders.

The net impact of the integration of safety, security, and preparedness is an increase in the overall effectiveness of the integrated response to any plant challenge. Over the lifetime of the plant design, operational, security, and preparedness changes are likely to be proposed. Such changes should be evaluated for their impact on integrated safety, security, and preparedness using the same security performance standards and integrated decision making process described above.

3.6   References

[NRC 2002]   U.S. Nuclear Regulatory Commission. Washington, D.C. "All Operating Power Reactor Licensees; Order Modifying Licenses (Effective Immediately), EA-02-26." *Federal Register:* Vol. 67, No. 42. pp. 9792–9796. March 4, 2002.

[NRC 2003]   U.S. Nuclear Regulatory Commission. Washington, D.C. "All Operating Power Reactor Licensees; Order Modifying Licenses (Effective Immediately), EA-03-086." *Federal Register:* Vol. 68, No. 88. pp. 24510–24514. May 7, 2003.

[NRC 2007]   U.S. Nuclear Regulatory Commission. SECY-07-0167, "Revision of Policy Statement on Regulation of Advanced Reactors." NRC: Washington, D.C. September 25, 2007.

4. DEFENSE-IN-DEPTH: TREATMENT OF UNCERTAINTIES

4.1 Introduction

The purpose of this chapter is to describe the approach used for implementing defense-in-depth for future reactors. An approach to defense-in-depth is developed for future reactors that is consistent with the successful past practices used for operating reactors, but which improves on such practices by being more consistent and by using quantitative information where possible. As described in the remainder of this chapter, the Framework's defense-in-depth approach combines deterministic elements with probabilistic ones.

The core of the U.S. Nuclear Regulatory Commission's (NRC's) safety philosophy has always been the concept of defense-in-depth, and defense-in-depth remains basic to the safety and security expectations of the risk-informed, performance-based Framework. The ultimate purpose of defense-in-depth is to compensate for uncertainty. This includes uncertainty in the type and magnitude of challenges to safety, as well as in the measures taken to ensure safety. Figure 4-1 shows the relationship of defense-in-depth to the rest of the structure of the Framework.

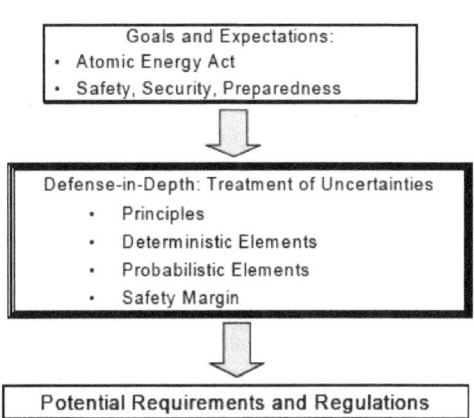

Figure 4-1   Framework Approach to Defense-in-Depth

4.1.1 Overview

The definition of defense-in-depth used in this document is the following: *Defense-in-depth is an element of NRC's safety philosophy that is used to address uncertainty by employing successive measures including safety margins to prevent and mitigate damage if a malfunction, accident or naturally caused event occurs at a nuclear facility.* As implied by this definition, to compensate for uncertainty a principal objective of defense-in-depth is to ensure that safety will not be wholly dependent on any single element of the design, construction, maintenance, or operation of a nuclear facility. This objective should always apply, except where the public health and safety consequences of the regulated activity and their uncertainties are small. This chapter discusses implementing defense-in-depth for future reactors and is based upon the following principles, which address the scope and approach taken for defense-in-depth:

1.   Measures against intentional acts as well as inadvertent events are provided.

2.   The design provides accident prevention and mitigation capability.

3.   Accomplishment of key safety functions is not dependent upon a single element of design, construction, maintenance or operation.

4.   Uncertainties in systems, structures, and components (SSCs) and human performance are accounted for in the safety analysis, and appropriate safety margins are provided.

5.   The plant design has containment functional capability to prevent an unacceptable release of radioactive material to the public.

6.   Plants are sited at locations that facilitate protection of public health and safety.

This chapter also addresses the most common defense-in-depth measures arising from these principles. These are providing redundancy, diversity, and safety margins, both in the equipment and the human actions that are important for the safe operation of the plant. Redundancy enhances the reliability of independent means to accomplish a needed function; diversity (and separation) generally provide protection against dependent (common cause) failures of multiple means. Allowances in excess of minimum requirements for physical parameters such as the capacities of hydraulic, electrical, and structural components contribute to safety margins that ensure unanticipated increases in demand can be met. Allowances in excess of minimum requirements for temporal parameters, such as time needed for operator actions and preventive systems to correct for deviations, contribute to safety margins that ensure deviations can be remedied even after some initial lapses. Consensus codes and standards is one of the means used to incorporate such allowances into design and operation, and thus contributes to safety margins, and therefore defense-in-depth.

As discussed in the rest of this chapter, the Framework approach to defense-in-depth incorporates both deterministic and probabilistic elements.

The two principal deterministic defense-in-depth elements of the approach are

(1)   ensuring the implementation of all of the five protective strategies introduced in Chapter 2 and discussed in detail in Chapter 5 (Physical Protection, Stable Operation, Protective Systems, Barrier Integrity, and Protective Actions). The protective strategies were selected based on engineering judgment, as a minimal set to provide protection for lines of defense against accidents and exposure of the public and environment to radioactive material. This set of protective strategies provides a high-level defense-in-depth structure.

(2)   ensuring that the defense-in-depth principles, discussed in Section 4.4, are followed to develop licensing potential requirements. As described in Section 4.3, the defense-in-depth principles are established by examining the different kinds of uncertainties to be treated, and incorporating successful past practices and lessons learned related to defense-in-depth.

The probabilistic elements of the approach consist of

(1)   using the Probabilistic Risk Assessment (PRA), to the extent possible, to search for and identify unexpected scenarios, including their associated uncertainties.

(2)   to subsequently establish adequate defense-in-depth measures, including safety margins, to compensate for those scenarios and their uncertainties which are quantified in the PRA model. The ability to quantify risk and estimate uncertainty using PRA techniques, where possible, and taking credit for defense-in-depth measures in risk analyses, allows one to provide a better estimate of how much defense-in-depth is enough. In this manner PRA complements defense-in-depth.

4.1.2   Background

The March 1999 Commission White Paper on risk-informed and performance-based regulation states that, *"Defense-in-depth is an element of the NRC's Safety Philosophy that employs successive compensatory measures to prevent accidents or mitigate damage if a malfunction, accident or naturally caused event occurs at a nuclear facility."* In its discussion on risk-informed approach and defense-in-depth the White Paper further states, "Although uncertainties associated with the importance of some elements of defense may be substantial, the fact that these elements and uncertainties have been quantified can aid in determining how much defense makes regulatory sense."

Regulatory Guide 1.174, which deals with risk-informed decision making on changes to the licensing basis of plants, states that "... the staff expects that:.....appropriate consideration of uncertainty is given in analyses and interpretation of findings, including using a program of monitoring, feedback, and corrective action to address significant uncertainties." It further states that "Defense-in-depth... has been and continues to be an effective way to account for uncertainties in equipment and human performance. If a comprehensive risk analysis is done, it can be used to help determine the appropriate extent of defense in depth (e.g., balance among core damage prevention, containment failure, and consequence mitigation) to ensure protection of public health and safety."

While Regulatory Guide 1.174 and other references always link defense-in-depth and safety margin,[9] the terms are often discussed separately. The Framework definition of defense-in-depth includes safety margin as an integral part.

Safety analysts [Weick 2001] pointed out that the key to safe operations in any activity is a focus on managing the "unexpected." The concept of defense-in-depth is essential for successfully coping with unexpected and uncertain events. Managing the unexpected includes identifying, evaluating, and managing uncertainties. In licensing future reactors, the treatment of uncertainties will play a key role in ensuring safety limits are met and the design is robust for unanticipated factors. In general, uncertainties associated with future plants will tend to be larger than uncertainties associated with existing plants due to new technologies being used, the lack of operating experience or, in the case of some proposed plants, new design features (e.g., increased use of passive systems). Any licensing approach for future plants needs to account for the treatment of these uncertainties.

Uncertainties have always been a factor to contend with in the safe operation of nuclear power plants, and, as stated in Regulatory Guide 1.174, "The defense in depth philosophy ......has been and continues to be an effective way to account for uncertainties in equipment and human performance." Note, however, that the defense-in-depth discussed in Regulatory Guide 1.174 is focused on currently operating plants, where defense-in-depth has been well established and confirmed by extensive operating experience. Regulatory Guide 1.174 is primarily concerned with maintaining defense-in-depth when contemplating changes to the licensing basis of an existing plant. For future reactors the challenge is somewhat different: Establishing what constitutes an adequate level of defense-in-depth in the future reactor designs. In laying the groundwork for defense-in-depth for future reactor designs, analysts can benefit from the experience of past designs, but at the same time need to be willing to re-examine conclusions based on existing designs and consider new alternatives.

---

[9] For example, Reg Guide 1.174 lists among key principles to be met that: (1) the proposed change is consistent with the defense-in-depth philosophy, and (2) the proposed change maintains sufficient safety margins.

Probabilistic Risk Assessment (PRA) is a powerful tool for searching for the unexpected and identifying uncertainties. Its original purpose was exactly this kind of search and identification. Much of the work of PRA for future reactors will be to identify and evaluate initially unexpected scenarios. While PRA cannot compensate for the unknown and identify all unexpected events or event sequences, it can (1) identify some originally unforseen scenarios, (2) identify where some of the uncertainties lie in the plant design and operation, and, for some uncertainties, (3) quantify the extent of the uncertainty, and therefore lead to a safer plant design. Consequently, PRA has a role to play, along with deterministic considerations, in establishing what constitutes adequate defense-in-depth.

PRA can quantify parameter uncertainty associated with the basic data used in the plant model. It can also address, to some extent, the model uncertainty associated with the analytical physical models and success criteria that appear because of modeling choices. For those uncertainties that are able to be quantified, PRA can indicate how much defense-in-depth, including margin, is needed to compensate for uncertainty to ensure safety.

Uncertainty associated with limitations in knowledge, such as unknown or unforeseen failure mechanisms, or unanticipated physical and chemical interactions among system materials, cannot be identified by PRA. Defense-in-depth measures to address this type of uncertainty can be established from repeatedly asking the question, "What if this barrier, measure, or safety feature fails?" without a quantitative estimate of the likelihood of such a failure, as well as by ensuring consistency with established defense-in-depth principles. This approach to defense-in-depth invokes specific deterministic provisions to compensate for the unexpected.

One approach (Sorenson 1999) classified defense-in-depth into two basic types: a more or less deterministic approach, referred to as the structuralist approach, and an approach that includes probabilistic assessments of uncertainty, the rationalist approach. According to the deterministic model defense-in-depth is embodied in the structure of the regulations and in the design of the facilities that are built in accordance with those regulations. The potential requirements for defense-in-depth result from repeatedly asking the question, "What if this barrier or safety feature fails?" regardless of the quantitative estimate of the likelihood of such a failure. Therefore, a characteristic of this approach is that some reliance on each of the lines of defense is maintained; accident prevention alone cannot be relied on to reach an acceptable level of safety. This approach to defense-in-depth has traditionally been used to achieve adequate protection. The elements of this deterministic, or structuralist, approach address primarily completeness uncertainties.

In the probabilistic model, defense-in-depth is the aggregate of provisions made to compensate for uncertainty and incompleteness in our knowledge of accident initiation and progression. The probabilistic approach acknowledges PRA as a powerful tool in searching for the unexpected and identifying uncertainties. Although PRA cannot compensate for the unknown and identify all unexpected events, this approach uses risk assessment to:

(1)   identify some originally unforseen scenarios,

(2)   identify where some of the uncertainties lie in the plant design and operation, and, for some uncertainties,

(3)   quantify the extent of the uncertainty.

This approach recognizes that PRA can identify and quantify parameter uncertainty associated with the basic data used in the plant model, and can also address, to some extent, the model uncertainty associated with the analytical physical models and success criteria that appear because of modeling choices. The probabilistic approach seeks to evaluate the uncertainties in the analysis and to determine what steps are to be taken to compensate for those uncertainties. The adequacy of the defense-in-depth measures can be assessed in the probabilistic approach via quantitative criteria that appear in safety goals or more general frequency/consequence curves. The elements of the probabilistic approach address primarily modeling and parameter uncertainties and allow an estimate of how much defense-in-depth, including margin, is needed in these areas.

4.2   Types of Uncertainty

Uncertainties have generally been categorized into random, or stochastic uncertainty (sometimes referred to as aleatory) and state-of-knowledge uncertainty (sometimes referred to as epistemic [Sorenson 1999]. Random uncertainty arises from events or phenomena occurring randomly or stochastically, such as a pump failing to start due to a random failure. Random uncertainty is sometimes called irreducible uncertainty because, in principle, it cannot be further reduced by additional empirical studies. However, additional study may lead to a better characterization, for example in terms of its magnitude, of the random uncertainty. Random uncertainty is well suited to analysis via probability theory, and this type of uncertainty is usually addressed in PRAs because it is embedded within the structure of the probabilistic models used to describe the occurrences of these events.

State-of-knowledge uncertainty arises from a lack of knowledge or lack of scientific understanding that may be due to a variety of factors, such as the inability to make observations, measurement uncertainty, the prohibitive cost of investigating a phenomena. State-of-knowledge uncertainty can be reduced, at least in principle, by additional study (theoretical research, experiments) or improved study techniques. Random and state-of-knowledge uncertainties are often intertwined and may be difficult to distinguish: measurement uncertainty usually has a random component; some apparent randomness may prove to be state-of-knowledge after closer examination. The state-of-knowledge uncertainties that need to be accounted for in a PRA fall into three basic categories:

*   *Parameter uncertainty* is the uncertainty associated with basic data used in safety analysis such as failure rates, ultimate strength. Part of parameter uncertainty is already included within random uncertainty, such as the beta or error factor; however, another part, such as the limitations in data affecting the choice of failure distribution, may be characterized as state-of-knowledge uncertainty. Parameter uncertainties are those associated with the values of parameters of the PRA models. (A pump that may or may not start is a random process, while determining the values to assign to the probability model for that failure event is a state-of-knowledge uncertainty.) Parameter uncertainties are typically characterized by establishing probability distributions on the parameter values. These distributions can be interpreted as expressing a degree of belief in the values these parameters could take, based on current knowledge and conditional on the underlying model being correct.

*   *Model uncertainty* is the uncertainty associated with the data limitations, analytical physical models, and acceptance criteria used in the safety analysis. PRA models, as well as those used in traditional deterministic engineering analyses, are composed of models for specific events or phenomena. Often the state of knowledge for these events and phenomena is incomplete, and expert opinions vary on how to formulate particular models. Such uncertainties arise, for example, in modeling human performance; common cause failures;

mechanistic failures of structures, systems and components; high-temperature fuel phenomena; and large radionuclide releases. While some model uncertainties will apply over several technologies, each particular technology will have its own special model uncertainties. Therefore, model uncertainties have to be identified at the technology-specific level as well. Model uncertainties are large where phenomena are poorly understood or not well characterized. Understanding the model uncertainties inherent in a particular PRA prediction for any future reactor design and how they are treated in terms of the available defense-in-depth elements is important.

- ***Completeness uncertainty*** is the uncertainty associated with factors not accounted for in the safety analysis such as safety culture, unknown or unanticipated failure mechanisms. Completeness uncertainty can be regarded as one aspect of modeling uncertainty, but because of its importance, is usually discussed separately. In one sense, it can be considered a scope limitation. Because completeness uncertainty reflects the unanalyzed contribution to risk it is difficult to estimate its magnitude, which can translate to difficulties estimating the true magnitude of the overall risk. Completeness uncertainty refers to things that are not modeled because of

    (1) _Intentional exclusion from the scope_. This includes risk contributors that can be modeled but are excluded for reasons of time, cost, etc., or a belief that their risk contribution for the analysis performed is negligible or can be adequately bounded. Some prominent examples are the risk from external hazards that are known to be extremely small, or accidents at low power, and shutdown for some plant-specific analyses.

    (2) _Lack of knowledge_. This consists of the truly unknown and unexpected that remains after available (and practical) analytical and experimental methods have been exhausted. This uncertainty is made up of a) risk contributors (e.g., initiating events and accident scenarios) that have not been conceived, and b) considerations for which adequate methods of analysis have not been developed, for example, heroic acts and influences of organizational performance. This type of uncertainty is most difficult to address in terms of what is adequate defense-in-depth. As noted in the Introduction to this chapter, defense-in-depth measures to address this type of uncertainty cannot be established via specific deterministic or probabilistic analysis, but instead rely on adherence to well thought-out defense-in-depth principles and from repeatedly asking the question, "What if this barrier or safety feature fails?"

4.3   Defense-in-Depth Objectives and Principles

As stated in the introduction to this chapter, the ultimate purpose of defense-in-depth is to compensate for uncertainty. All of the uncertainties described above can arise in the analysis of the challenges to safe operation, and in the design of actions and equipment to ensure safety. As noted, uncertainties related to lack of knowledge are the most difficult to deal with. Based on these considerations, the purpose of defense-in-depth can be expressed with the objectives shown below. Defense-in-depth is the ability to

- compensate for uncertainties, including events and event sequences which are unexpected because their existence remained unknown during the design phase,

- compensate for potential adverse equipment performance, as well as human actions of commission (intentional adverse acts are part of this) as well as omission,

- maintain the effectiveness of barriers and protective systems by ensuring multiple, generally independent and separate, means of accomplishing their functions, and

- protect the public and environment if these barriers are not fully effective.

The first objective emphasizes the importance of providing some means to counterbalance unexpected challenges. The second objective addresses uncertainty in equipment and human actions. It encompasses equipment design and fabrication errors, as well as both deliberate acts meant to compromise safety, and errors or inadequacy in carrying out procedures meant to ensure safety. The third objective addresses the uncertainty in the performance of the systems, structures, and components (SSCs) that constitute the barriers to radionuclide release, as well as in the SSCs whose function is to protect those barriers. The final objective emphasizes the concept of layers of protection, in that it addresses the need for additional measures should the barriers to radionuclide release fail after all.

Much can be learned from the successful past applications of defense-in-depth. The most well known is the use of multiple physical barriers, exemplified in current reactors by the fuel elements and cladding, reactor coolant system pressure boundary and containment systems and structure to prevent the release of significant quantities of radionuclides to the environment. The application here has also included the design of redundant and diverse independent active and passive systems which protect the integrity of these barriers.

Recurrent themes in applications of defense-in-depth are (1) do not rely on one element of design no matter how confident, and (2) guard against the unexpected, i.e., don't assume accidents will start and play out in the analyzed way. These themes of defense-in-depth have been applied in various ways. Redundant or diverse, generally independent means are usually used to accomplish key safety functions, such as safe shutdown or removal of decay heat. Redundancy enhances the reliability of independent means; diversity (and separation) generally provides protection against dependent (common cause) failures of multiple means, and therefore some assurance that safety functions can be accomplished successfully despite the uncertainty in the mechanism of dependent failures. In some advanced designs, additional emphasis is given to inherent reactor characteristics and passive features that minimize the potential for radionuclide release and reduce barrier failure modes, even for unanticipated accident scenarios, as ways of assuring safety functions are accomplished. In these designs safety functions may be achieved by inherent natural processes such as shutdown due to negative reactivity feedback, or decay heat removal through conduction and radiation to surrounding structures, and retention of fission products in high integrity fuel particles.

Defense-in-depth measures have been embodied in SSCs, in procedures (including accident management plans to protect the off-site public), or in the choice and design of the basic processes that promote safety (e.g., negative temperature coefficient of reactivity).

Based on these past defense-in-depth practices and lessons learned from operating experience, security assessments, and the consideration of the various uncertainties that are to be dealt with, a set of defense-in-depth principles were established.   To ensure   public safety despite uncertainties in knowledge or rigor, the first general principle of defense-in-depth is that:

(1)   ***Measures against intentional as well as inadvertent events are provided.***

This principle ensures that defense-in-depth measures are applied not just against random failures of SSCs or human errors, but also against acts of sabotage, theft of nuclear materials, armed intrusion, and external attack. Such measures can be incorporated in the design of the plant, be part of operating practices, and include the capability to respond to intrusion or attack. This principle then calls for defense-in-depth considerations to be applied to all types of plant disturbances: internal events arising from random equipment and human failure; to external events resulting from earthquakes, fires, floods, high winds, etc.; and intentional destructive acts such as sabotage, diversion, and attack. The importance of including defense-in-depth in physical protection measures that address deliberate destructive acts is increased by the fact that physical protection affects all the other protective strategies.

Past discussions of defense-in-depth, at least implicitly, focused primarily on the application of defense-in-depth to compensate for potential human errors, and component failures arising from 'inadvertent' causes such as aging, corrosive processes, poor design. However, with the increased need to consider security issues, embodied in the protective strategy of physical protection, defense-in-depth considerations also include protection against intentional acts directed at nuclear plants that would threaten public health and safety.

This principle ensures that defense-in-depth is considered when implementing physical protection measures, and therefore also implies that the subsequent principles listed here are invoked for physical protection measures just as they are invoked for measures used to achieve the other protective strategies. For example, the strategy of physical protection calls for both preventive and mitigative measures. This is in keeping with conventional approaches to security. For future reactors physical protection measures can be considered during the design stage via vulnerability assessments at this stage. In this manner these measures can become integral with the design and the mitigative and preventive features thus applied are likely to be better than features added as an afterthought to the design.

From this first general principle of defense-in-depth, five additional defense-in-depth principles follow:

(2)   ***The design provides accident prevention and mitigation capability.***

This principle ensures an apportionment in the plant's capabilities between limiting disturbances to the plant and mitigating them, should they occur. This apportionment is present in both the design and operation of the plant. It is not meant to imply an equal apportionment of capabilities. The protective strategies introduced in Chapter 2 provide an important illustration of this principle. Some of these strategies (stable operation, protective systems) are more preventive, while others (protective actions, and to some extent barrier integrity) are more mitigative. Physical protection clearly falls into both areas. By requiring that all of the strategies have to be incorporated into plant design and operation, the presence and availability of both preventive and mitigative features is ensured. The strategies are not 'equal' in terms of their contribution to quantitative risk reduction, for example, but none are completely absent from the design and operation of the plant.

In general terms accident prevention can be defined as the measures used to prevent the uncontrolled migration of radionuclides within the plant in excess of normal operating limits. Accident mitigation can be defined as measures used to prevent the uncontrolled migration of radionuclides from the plant to the environment in excess of normal operational limits. In these definitions measures refer to SSCs and procedures, and the plant refers to the physical structures that create a boundary between the environment and any sources of radioactivity at the site. (Appendix C of this report further explores the issue of defining prevention and mitigation generically.) Reducing the frequency of initiating events is generally viewed as a preventive measure; if the initiating events occurs, then helping to cope with its consequences is seen as a mitigative measure. A given system, structure or component may, in fact, serve to prevent one challenge and mitigate another challenge, depending on where it occurs in an event sequence. Often prevention is emphasized relative to mitigation for a variety of reasons. Preventive measures are usually more economical; prevention avoids having to deal with the phenomenological uncertainties that arise once an accident progresses From a defense-in-depth standpoint such an emphasis is acceptable as long as it does not result in an exclusive reliance on prevention with a total neglect of mitigative features.

For both commercial and safety reasons, there is likely to be a great deal of emphasis on the protective strategy of stable operation. Such an approach tries to prevent deviations from normal operation, and system failures. For intentional events, the physical protection strategy will also have as its dominant focus the limitation of initiating events resulting from such acts.

The next protective strategy, ensuring that protective systems are available, recognizes that some initiating events are likely to occur over the service lifetime of a nuclear power plant, despite the care taken to prevent them. This strategy has a preventive component in that some of these systems are concerned with detecting and intercepting deviations from normal operation to prevent anticipated operational occurrences from escalating to accident conditions. However, protective systems also include systems that play a dual role of prevention and mitigation or a strictly mitigative role. In practice, safety systems will likely be used for both aspects of defense. This aspect of the protective system strategy recognizes that, although very unlikely, the escalation of certain anticipated operational occurrences or other initiating events may not be arrested, and a more serious event may develop. These unlikely events are anticipated in the design basis for the plant, and inherent safety features, as well as additional equipment and procedures are likely to be provided to control their consequences and to achieve stable and acceptable plant states following such events. This leads to the need that engineered safety features are provided that are capable of leading the plant first to a controlled state, and then to a safe shutdown.

The strategy of barrier integrity plays mainly a mitigative role. While the barrier associated with the fuel prevents an off-normal event from escalating, successive barriers mitigate the consequences of the failure of the fuel barrier. The latter barriers often include the protection offered by a containment or confinement, but may also be achieved by complementary measures and procedures to arrest accident progression, and by mitigating the consequences of selected severe accidents. Adequate excess capacities in the equipment, structure, and procedures used here to provide safety margins are an important part of the strategy. The physical protection strategy may also introduce barriers against internal and external threats that could compromise plant safety systems.

The increased use of inherent safety characteristics and passive features could strengthen accident prevention as well as mitigation in new and innovative designs.

The strategy of protective actions, such as accident management, is purely mitigative. This strategy includes accident management procedures within the plant (for which margins in barrier strength and in the time needed to achieve successful accident management are essential), as well as emergency response. The emergency response part is aimed at mitigating consequences of potential releases of radioactive materials that may result from accidents. Such an emergency response requires an adequately equipped emergency control center, and plans for on-site and off-site emergency response to protect the public's health and safety. These plans should include ensuring that emergency planning is consistent with plant safety by accounting for factors such as plant risk, timing, form, magnitude and duration of releases of radioactive material, and dose to the public as a function of distance from the plant. Sufficient allowances in temporal parameters to ensure safety margins are important considerations. Physical protection aspects introduce additional considerations into both on-site and off-site protective actions.

(3)   ***Accomplishment of key safety functions is not dependent upon a single element of design, construction, maintenance or operation.***

This principle ensures that redundancy, diversity, and independence in SSCs and actions are incorporated in the plant design and operation, so that no key safety functions will depend on a single element (i.e., SSC or action) of design, construction, maintenance or operation. The key safety functions include (1) control of reactivity, (2) removal of decay heat, and the functionality of physical barriers to prevent the release of radioactive materials. In addition, hazards such as fire, flooding, seismic events, and deliberate attacks, which have the potential to defeat redundancy, diversity, and independence, need to be considered.

An important aspect of ensuring that key safety functions do not depend on a single element of design, construction, or operation is guarding against common cause failures and consequential failures. Failure of several devices or components to function may occur as a result of a single specific event or cause. Such failures may affect several different items important to safety simultaneously. The event or cause may be a design deficiency, a manufacturing deficiency, an operating or maintenance error, a natural phenomenon, a human-induced event (intentional or inadvertent), or an unintended cascading effect from any other operation or failure within the plant. Common cause failures may also occur when several of the same type of components fail at the same time. This failure may be due to reasons such as a change in ambient conditions, repeated maintenance error or design deficiency. Measures to minimize the effects of common cause failures, such as the application of redundancy, diversity and independence, are an essential aspect of defense-in-depth. Redundancy, the use of more than a minimum number of sets of equipment to fulfill a given safety function, is an important design principle for achieving high reliability in systems important to safety. Redundancy enables failure or unavailability of at least one set of equipment to be tolerated without loss of the function. For example, three or four pumps might be provided for a particular function when any two would be capable of satisfying the specified acceptance criteria. For the purposes of redundancy, identical or diverse components may be used. Consequential failures may occur as a result of high-energy line breaks, radiation damage, and structural failures.

The reliability of some safety functions can be improved by using the principle of diversity to reduce the potential for certain common cause failures. Diversity is applied to redundant systems or components that perform the same safety function by incorporating different attributes into the systems or components. Such attributes could be different principles of operation, different physical variables, different conditions of operation or production by different manufacturers, for example.

To ensure diversity is actually achieved, the designer examines some of the more subtle aspects of the equipment used. For example, to reduce the potential for common cause failures, the designer examines the application of diversity for any similarity in materials, components and manufacturing processes, or subtle similarities in operating principles or common support features. In addition, if diverse components or systems are used, there is a reasonable assurance that such additions are of overall benefit, i.e., reliability is actually improved, taking into account the disadvantages such as the extra complication in operation, maintenance and testing, or the consequent use of equipment of lower reliability.

Another important aspect of this defense-in-depth principle is the use of functional isolation, physical separation, and physical protection to achieve independence among safety systems. The reliability of plant systems can be improved by maintaining the following features for independence in design:

- independence among redundant system components;

- independence between system components and the effects of certain initiating events such that, for example, an initiating event does not cause the failure or loss of a safety system or safety function that is necessary to mitigate the consequences of that event;

- independence between or among systems or components of different safety classes; and

- independence between items important to safety and those not important to safety.

Functional isolation can be used to reduce the likelihood of adverse interaction between equipment and components of redundant or connected systems resulting from normal or abnormal operation or failure of any component in the systems.

Physical separation in system layout and design can be used to increase assurance that independence will be achieved, particularly in relation to certain common cause failures, including deliberate acts intended to defeat safety systems.

Physical separation and physical protection by design includes separation by
- location (such as distance or orientation);
- barriers; or
- a combination of these.

The means of separation will depend on the challenges considered in the design basis, such as effects of fire, chemical explosion, aircraft crash, missile impact, flooding, extreme temperature, humidity, or radiation level, as well as deliberate acts of disabling or destroying safety systems. Certain areas of the plant naturally tend to be centers where equipment or wiring of various levels of importance to safety will converge. Examples of such locations for Light Water Reactors (LWRs) may be containment penetrations, motor control centers, cable spreading rooms, equipment rooms, the control room and the plant process computers. These locations are particularly scrutinized and measures are taken to avoid common cause and consequential failures.

Functional isolation and physical separation are also likely to be important considerations for achieving adequate physical protection measures. 'Pinch points' in terms of functional performance as well as physical location can lead to vulnerabilities resulting from either accidental or intentional events.

Finally, this principle also requires that measures are included in the design and operation so that catastrophic events, such as an initiating event that prevents all safety features from operating, for example, are of low enough frequency that they do not have to be considered in the analysis. Examples of such events are pressurized thermal shock (in current reactors) that leads to catastrophic reactor vessel failure, or earthquakes beyond the design basis, but can also include deliberate attacks against the plant.

(4) ***Uncertainties in SSCs and human performance are accounted for in the safety analysis and appropriate safety margins are provided.***

This principle ensures that when risk and reliability goals are set, at the high level and the supporting intermediate levels, the design and operational means of achieving these goals account for the quantifiable uncertainties, and provide some measure of protection against the ones that cannot be quantified as well.

When allocating risk goals that meet the overall risk criteria, a designer includes allowances for uncertainty. For example, a designer can allocate reliability targets for each of the protective strategies introduced in Chapter 2. These targets will be supported by maximum unavailability limits for certain safety systems to ensure the necessary reliability for the performance of safety functions and the strategies. Uncertainties are factored into the establishment of all these targets.

An important tool for achieving risk goals for design, construction, and operation of the plant is the use of risk assessments that include estimates of uncertainty. The setting of success criteria for achieving safety functions are established, and the calculations that show they have been met are performed in such a way that uncertainties are accounted for with a high level of confidence. At least initially, this approach needs to be done for future reactor designs without always having the benefit of reviewing past performance.

Ensuring adequate safety margins is important here in achieving a robust design. (Safety margins are further discussed in Section 4.6.) Excess capacity in physical and temporal parameters are incorporated in the plant equipment and procedures. Allowances in excess of minimum requirements for physical parameters such as the capacities of hydraulic, electrical, and structural components contribute to safety margins that ensure unanticipated increases in demand can be met. Allowances in excess of minimum requirements for temporal parameters, such as time needed for operator actions and preventive systems to correct for deviations, contribute to safety margins that ensure deviations can be remedied even after some initial lapses. Therefore, careful attention is paid to selecting design codes and materials, and to controlling fabrication of components and of plant construction. In addition, performance monitoring and feedback is used over the life of the plant to ensure reliability and risk goals continue to be met, or if not, corrective actions are taken.

Some future reactor designs use passive systems and inherent physical characteristics (confirmed by sensitive non-linear dynamical calculations and safety demonstration tests) to ensure safety, rather than relying on the performance of active electrical and mechanical systems. For such plants, with many passive systems, fault trees may be very simple when events proceed as expected and event sequences may have very low frequency and little apparent uncertainty. The real work of PRA for these designs may lie in searching for unexpected scenarios and their associated uncertainties, including unexpected safety system performance. Innovative ways to structure the search for unexpected conditions that can challenge design assumptions and passive system performance will need to be developed or identified and applied to these facilities. The risk may arise from unexpected ways the facility can end up operating outside the design assumptions. For example, a hazard and operability (HAZOP) related search scheme for scenarios that deviate from designers' expectations and a structured search for construction errors and aging problems may be the appropriate tools. Examples of uncertainties in design and operation that can lead to ways in which the facility can operate outside its design assumptions include scenarios where:

- the human operators and maintenance personnel place the facility in unexpected conditions;

- gradual degradation has led to unobserved corrosion or fatigue or other physical condition far from that envisioned in the design; or

- passive system behavior (e.g., physical, chemical, and material properties) is incorrectly modeled.

Measures used for physical protection are designed to account for uncertainties as well. Security assessments for future reactors include some considerations of beyond design basis threats (DBTs) to address uncertainties.

(5) *The plant design has containment functional capability to prevent an unacceptable release of radioactive material to the public.*

This principle ensures that regardless of the features incorporated in the plant to prevent an unacceptable release of radioactive material from the fuel and the reactor coolant system (RCS), there are additional means to prevent an unacceptable release to the public should such a release occur that has the potential to exceed the dose acceptance criteria. The purpose of this principle is to protect against unknown phenomena and threats, i.e., to compensate for completeness uncertainty affecting the magnitude of the source term.

The containment for preventing unacceptable radionuclide releases to the environs should ensure that the design has adequate capability to reduce radionuclide release to meet the on-site and off-site radionuclide dose acceptance criteria. In doing so, threats from selected low probability, but credible events, with the potential for a large source term and a significant radionuclide release to the environs are also considered.

Adequate data are required to provide the quantitative basis for the performance of each of the mechanistic barriers and obstacles for the range of plant conditions associated with the selected events in each category. For future reactor technologies it will be difficult to ensure that complete data, spanning all credible events, will be available. Therefore, even if the mechanistic source term calculations indicate that releases from the fuel and RCS are small enough to meet release criteria, other means need to be available to prevent uncontrolled releases to the environment. These means will also be important for threats that are addressed under physical protection. Accordingly, each design needs to have the capability to establish a controlled low leakage barrier if plant conditions result in the release of radioactive material from the fuel and the reactor coolant system in excess of anticipated conditions. The specific conditions for the barrier leak tightness, temperature, pressure, and time available to establish the low leakage condition will be design-specific. The design of the controlled leakage barrier should be based upon a process that defines a hypothetical event representing a serious challenge to fission product retention in the fuel and the coolant system. The applicant and NRC should agree upon a hypothetical event consistent with the technology and safety characteristics of the design. (Chapter 8 provides additional details on analyzing such an event to demonstrate that this defense-in-depth principle is satisfied.) As noted above, the particular means used to retain or control the release will depend on the reactor technology.

(6)   ***Plants are sited at locations that facilitate the protection of public health and safety.***

This principle ensures that the location of regulated facilities facilitates the protection of public health and safety by considering population densities and the proximity of natural and human-made hazards in the siting of plants. Physical protection aspects associated with security concerns are additional considerations in selecting the site. Siting factors and criteria are important in ensuring that radiological doses from normal operation and postulated accidents will be acceptably low, that natural phenomena and potential human-made hazards will be accounted for in the design of the plant, that site characteristics are such that adequate security measures to protect the plant can be developed, and that physical characteristics unique to the proposed site that could pose a significant impediment to developing emergency plans are identified. The safety issues to be considered include geologic/seismic, hydrologic, and meteorological characteristics of proposed sites; exclusion area and low population zone; population considerations as they relate to protecting the general public from the potential hazards of serious accidents; potential effects on a station from accidents associated with nearby industrial, transportation, and military facilities; emergency planning; and security plans. The environmental issues to be considered concern potential affects from the construction and operation of nuclear power stations on ecological systems, water use, land use, the atmosphere, aesthetics, and socio-economics.

For reactors, this principle is also intended to ensure that protective actions, including emergency preparedness (EP), are a fundamental element of defense-in-depth. EP includes an emergency plan that provides notification, drills, training, sheltering, and evacuation.

These defense-in-depth principles are based upon and consistent with the Commission's white paper, quoted earlier, that states defense-in-depth is: (1) an element of the NRC's Safety Philosophy that uses successive compensatory measures to prevent accidents or mitigate damage if a malfunction or accident occurs at a nuclear facility and (2) ensures that safety functions will not depend wholly on any single element of the design, construction, maintenance, or operation of a nuclear facility. The net effect of incorporating defense-in-depth into design, construction, maintenance, and operation is that the facility or system in question tends to be more tolerant of failures and external challenges. The principles are also consistent with Regulatory Guide 1.174, which states that consistency with the defense-in-depth philosophy is maintained if

- a reasonable balance is preserved among prevention of core damage, prevention of containment failure, and consequence mitigation.

- over-reliance on programmatic activities to compensate for weaknesses in plant design is avoided.

- system redundancy, independence, and diversity are preserved commensurate with the expected frequency, consequences of challenges to the system, and uncertainties (e.g., no risk outliers).

- defenses against potential common cause failures are preserved, and the potential for the introduction of new common cause failure mechanisms is assessed.
- independence of barriers is not degraded.

- defenses against human errors are preserved.

- the intent of the General Design Criteria in Appendix A to Part 50 of Title 10 of the Code of Federal Regulations (10 CFR Part 50) is maintained.

These points in Regulatory Guide 1.174 line up well with the defense-in-depth principles stated previously.

4.4   Defense-in-Depth Approach

As noted in the Introduction, the defense-in-depth approach used here is one that combines deterministic and probabilistic elements. The probabilistic elements are used to determine how much defense-in-depth is needed to compensate for the uncertainties that can be quantified. The deterministic elements compensate for the unquantified uncertainties, especially the unexpected threats resulting from completeness uncertainty.

The probabilistic aspects of the approach are the use of a PRA that includes in its calculations the uncertainty associated with the parameter values and models used in the PRA. The PRA is used ultimately to verify that the quantifiable margins and other defense-in-depth measures in the design make the quantified uncertainty range acceptable. The principal deterministic elements consist of ensuring the implementation of all of the five protective strategies introduced in Chapter 2, and ensuring that the defense-in-depth principles of Section 4.3 are implemented in the design and operation of the plant. The Requirements Development Process of Chapter 8 describes the formulation of requirements based on each of the protective strategies. An essential part of the process is the application of the defense-in-depth principles to each protective strategy.

Propagating the uncertainty distributions of the parameter values and models used in the PRA throughout the calculations provides a designer with estimates of the probability ranges of the modeled challenges to the plant. It also provides the probability ranges associated with the capabilities of the SSCs and procedures that address these challenges. During the design stage an iterative process is likely where the designer adds or modifies SSCs or procedures to achieve reliability goals for their capability that adequately cover the uncertainty ranges of the challenges. The final design will have adequate margins and redundancy in SSCs and procedures to respond acceptably to identified challenges, including their uncertainty, and to ultimately make the total risk acceptable. This process is how probabilistic aspects of the approach are used to determine how much defense-in-depth is needed to achieve the desired quantitative goals to address the uncertainty that can be quantified in the risk assessment.

The deterministic aspects of the defense-in-depth approach are embodied first of all in applying the entire combination of the protective strategies to the design. The objective of these strategies are restated here:

- The **Physical Protection** objective is to protect workers and the public against intentional acts (e.g., attack, sabotage, and theft) that could compromise the safety of the plant or lead to radiological release.

- The **Stable Operation** objective is to limit the frequency of events that can upset plant stability and challenge safety functions, during all plant operating states, i.e., full power, shutdown, and transitional states.

- The **Protective Systems** objective is to ensure that the systems that mitigate initiating events are adequately designed, and perform adequately, in terms of reliability and capability, to satisfy the design assumptions on accident prevention and mitigation during all states of reactor operation. Human actions to assist these systems and protect the barriers are included here.

- The **Barrier Integrity** objective is to ensure that there are adequate barriers to protect the public from accidental radionuclide releases from all sources. Adequate functional barriers need to be maintained to protect the public and workers from radiation associated with normal operation and shutdown modes and to limit the consequences of reactor accidents if they do occur. Barriers can include physical barriers as well as the physical and chemical form of the material that can inhibit its transport if physical barriers are breeched.

- The **Protective Actions** objective is to ensure that adequate protection of the public health and safety in a radiological emergency can be achieved should radionuclides penetrate the barriers designed to contain them. Measures include emergency procedures, accident management, and emergency preparedness.

Taken together the protective strategies are a classic example of the deterministic defense-in-depth approach that addresses completeness uncertainty. For example, what if stable operation cannot be maintained, as a result of either intentional or inadvertent acts? Protective systems will restore the plant to normal operation or limit the accident consequences. What if protective systems fail? Barriers will confine the radioactive material. What if barriers are degraded and allow fission products to escape? Protective actions will mitigate the consequences. Taken in this order of the operational sequence of events that would occur during an accident, the protective strategies take the form of a high-level defense-in-depth structure, as shown in Figure 4-2 below.

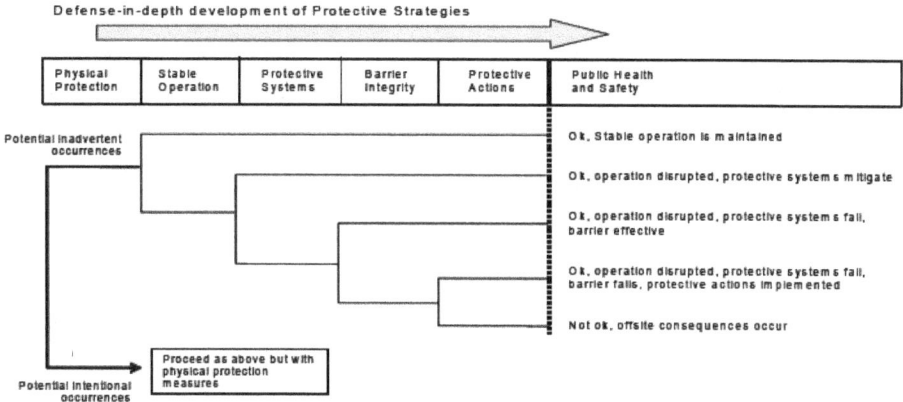

Figure 4-2   Protective Strategies as High-level Defense-in-depth

The figure also shows that physical protection supports all the other strategies, because if the upset condition is the result of an intentional act, then physical protection measures come into play along with the protective strategies.

Depending on the inherent characteristics of various new designs, the protective strategies may be accomplished by means substantially different from those used in the current light water reactors. Appendix A focuses on the safety characteristics of some of the new, innovative reactor designs, and how these inherent characteristics promote the success of the protective strategies, thereby contributing to defense-in-depth.

As stated above, an additional deterministic defense-in-depth aspect of the approach is the adherence to the defense-in-depth principles in implementing the individual protective strategies. This aspect is an essential part of the Requirements Development Process described in Chapter 8. All of the principles are met through the collective implementation of the protective strategies.

The intent of applying the defense-in-depth principles to each protective strategy is to ensure that defense-in-depth is considered in each line of defense, as well as in a broad sense across the entire design. The application of the defense-in-depth principles to the protective strategies to develop requirements is discussed in detail in Chapter 8. A brief summary of the application of the principles to each strategy is presented here.

Physical protection

- Physical protection considers intentional acts.

- Physical protection addresses prevention as well as mitigation. Considering security issues integral with the design process can lead to designs with enhanced prevention and mitigation features. Accordingly, a security assessment at the design stage should be performed.

- Physical protection need not depend upon a single element of design, construction or operation.

4. Defense-in-Depth

- Physical protection accounts for uncertainties.

- Since physical protection is directed toward preventing an unacceptable release of radioactive material to the environment, the security assessment includes an analysis of the release of radioactive material as a metric for decisions.

- Plant siting considers the ability to implement protective measures.

Stable operation

- Intentional acts to disrupt operation are considered, and such disruptions should be considered under physical protection.

- Designing the plant to prevent accidents is the main emphasis of the stable operation protective strategy. Chapter 6 of this document provides some measures to ensure that the assumptions in the PRA on initiating events are preserved.

- Event sequences considered in the design that could disrupt stable plant operation need not be of such a nature as to defeat the protective systems, barrier integrity, and protective actions strategies simultaneously.

- Uncertainties are considered in assessing the frequency of events that could disrupt stable plant operation. Accordingly, the PRA and safety analysis quantify uncertainties.

- Event sequences with the potential to defeat barrier integrity and protective actions strategies need to have a very low frequencies, as discussed in Chapter 6.

- The effect plant siting could have on contributing to the disruption of stable plant operation is considered in the design. This would include consideration of natural as well as man-made events.

Protective Systems

- At least some of the protective systems have the ability to respond to intentional acts as well as inadvertent events.

- Protective systems exist that prevent events from leading to major plant damage (prevention) as well as preventing the uncontrolled release of radioactive material to the environment should major plant damage occur (mitigation).

- Key plant safety functions (i.e., reactor shutdown and decay heat removal) are not dependent upon a single protective system. Accordingly, it is envisioned that each of those functions, be accomplished by redundant, independent and diverse means, with each means having reliability and availability goals commensurate with overall plant risk goals.

- In assessing the performance of protective systems, uncertainties in reliability, and performance are accounted for. For new types of equipment or equipment with little or no operating experience at the conditions it will experience, a reliability assurance program is provided to demonstrate and monitor equipment to ensure the assumptions of reliability, availability, and performance used in the PRA and safety analyses are met.

- The unacceptable release of radioactive material needs to be prevented. Accordingly, a means to prevent the uncontrolled release of radioactive material is included in the design, consistent with the barrier integrity protective strategy.

- Plant siting can affect the types and performance of safety systems since site-specific hazards may be different.

Barrier integrity

- The number of barriers and their design is based upon consideration of both intentional as well as inadvertent events.

- The barriers are designed with both accident prevention and mitigation in mind. Accident prevention is achieved by ensuring that the barriers are designed to be highly reliable and can withstand a range of off-normal conditions. Accident mitigation is achieved by ensuring the barriers perform their function of containing radioactive material.

- Defense-in-depth requires that key safety functions not depend upon a single element of design, construction, operation, or maintenance. Application of this principle to barrier integrity implies multiple barriers are used, since containment of radioactive material is considered a key safety function.

- In the design and safety analysis, uncertainties in reliability and performance are accounted for. However, not all uncertainties can be quantified. Therefore, it is considered reasonable to require each design to have additional capability to mitigate against accident scenarios that result in the release of larger amounts of radioactive material by providing margin to account for unquantified uncertainties that result in a larger source term available for release to the environment (e.g., security-related events).

- Barriers prevent the unacceptable release of radioactive material. Accordingly, to account for uncertainties, the design has a containment functional capability independent from the fuel and RCS.

- Barrier integrity interfaces with siting in that some aspects of barrier performance may be determined by site characteristics (e.g., meteorology, population distribution). Likewise, barrier integrity can also affect the type and extent of off-site protective measures.

Protective Actions

- The development of protective actions considers intentional acts as well as inadvertent events, and includes them in the physical protection protective strategy.

- Protective actions include measures to terminate the accident progression (referred to as emergency operating procedures, and accident management) and pre-planned measures to mitigate the accident consequences (referred to as emergency preparedness).

- The accomplishment of protective actions does not rely on a single element of design, construction, maintenance, or operation.

- Protective actions are developed in consideration of uncertainties. Emergency preparedness is included in the design and operation to account for unquantified uncertainties.

- Prevention of unacceptable releases of radioactive material is part of the protective action program.

- Plant siting affects protective actions, and is considered in developing plans.

It is also essential to restate here that, in applying defense-in-depth, security and preparedness are integral with safety. As stated in Chapter 3, an integrated and consistent approach to addressing safety, security, and preparedness is essential. Consideration of security in the design process is intended to result in a more robust, intrinsic security capability and rely less on extrinsic, operational security programs. Inclusion of preparedness in all aspects of safety and security integration leads to a plant that is safe and robust against all internal and external hazards and inadvertent and advertent threats. Finally, it should be noted that the degree of defense-in-depth can be different depending on the frequency of the challenges being assessed. This concept has been used in Chapter 6.

In summary, in the view of defense-in-depth used in this document a probabilistic approach to defense-in-depth is combined with a deterministic one. This concept has been incorporated and implemented in Chapters 5 and 6 of this report.

4.5   Safety Margin

Throughout this chapter safety margin has been described as an integral part of defense-in-depth, since the basic purpose of safety margins is to cope with uncertainty. For example, consensus codes and standards contribute to defense-in-depth because they incorporate safety margins. In addition, the compensatory measures that are also part of defense-in-depth need to have some margins embedded in them to deal with the accident and malfunction. Compensatory measures that lack margin are not very practical.

Although the term safety margin is frequently used in engineering design, there are different interpretations of the term "safety margin." The term is frequently found in regulatory documents that contain phrases such as "maintain adequate safety margin," or "provide sufficient safety margin," without a quantitative definition of safety margin. While this use of the term as a qualitative descriptor is useful in many contexts, for this document a more quantitative description is needed.

Safe operating conditions can be characterized by maintaining limits on one or more safety variables, such as pressure and temperature, etc. As stated in the draft NUREG-1870, "Integrating Risk and Safety Margins,"[10] safety margins are linked to safety limits—limiting values imposed on safety variables (e.g., peak clad temperature (PCT) and containment pressure in current LWRs). Thus, when operating conditions stay within safety limits, the safety barrier or system continues to function, and an adequate safety margin exists. The intent is to allow margin for phenomena and processes that are inadequately considered or neglected in the analysis predicting the behavior of the given system or physical barrier.

---

[10]NUREG-1870 is anticipated to be published in 2008.

For the definition of safety margin in this report, the safety variable is assumed to have an ultimate capacity, beyond which the safety system or barrier fails, e.g., the ultimate strength of a critical barrier. A regulatory limit is set on the safety variable, well below this capacity, to ensure that the ultimate capacity is not reached during normal operation as well as excursions from normal operation. The difference between the ultimate capacity and the regulatory limit is termed the "regulatory margin" here. The designer can incorporate an additional margin, called the "operational margin" here, by designing the system so it operates well below the regulatory limit for normal operations and excursions. Together the regulatory margin and the operational margin constitute the safety margin. Figure 4-3 shows this definition.

Figure 4-3 provides the background to a quantitative basis for safety margin, but is very simplified. In practice the capacity and the operational range of a plant safety variable are not single-valued quantities but have probability distributions associated with them. Questions immediately arise,

such as what measure of capacity should be used, what excursions from normal operation should be included, and what inadequately considered processes need to be compensated for. Defining margin in light of the probabilistic nature of the safety variables and the questions asked above are further pursued in Chapter 6.

From the above discussion providing adequate safety margin puts responsibility (1) on the regulator to place the regulatory limit with proper consideration of loads and capacities, and (2) on the designer to adequately set design limits to meet regulatory limits with 'margin.'

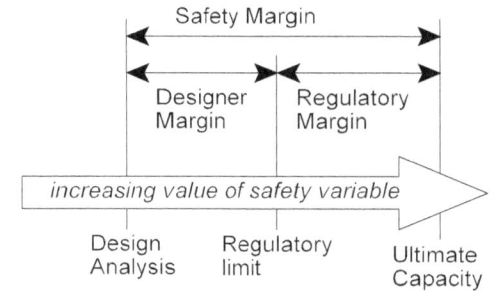

Figure 4-3   Safety Margin Definition

4.6   References

[Sorenson 1999]   Sorenson, J.N., G.E. Apostolakis, T.S. Kress, and D.A. Powers, "On the Role of Defense in Depth in Risk-Informed Regulation," presented at PSA '99, 1999.

[Weick 2001]   Weick, K.E., and K.M. Sutcliffe, "Managing the Unexpected: Assuring High Pressure in an Age of Complexity," Jossey-Bass, San Francisco, CA, 2001.

5. SAFETY FUNDAMENTALS: PROTECTIVE STRATEGIES

5.1 Introduction

The purpose of this chapter is to define the safety fundamentals that ensure protection of the public health and safety, and that accomplish the safety, security and preparedness goals and expectations. The safety fundamentals are defined in terms of five protective strategies. This chapter explains why these protective strategies are a sufficient set of safety fundamentals, and describes how they are used to develop potential requirements.

What overall process should be used to identify potential requirements for design, construction, and operation? Should the process to identify the potential requirements and regulations start with design considerations? Or, should the process start with some other consideration?

The process chosen to initially identify and define the potential requirements and regulations implements the safety, security, and preparedness expectations and ensures protection of the public health and safety. Safety fundamentals have been defined in the form of protective strategies that, if met, will ensure the protection of the public health and safety with a high degree of confidence.

The protective strategies approach is based on a philosophy of regulation that requires multiple strategies to ensure that there is little chance of endangering public health and safety. It is a top-down, hierarchical approach that starts with a desired outcome and identifies protective strategies (safety fundamentals) to ensure this outcome is achieved even if some strategies should fail. The protective strategies provide defense-in-depth that offer multiple layers of protection of public health and safety.

The five protective strategies introduced in Chapter 2 – *Physical Protection, Stable Operation, Protective Systems, Barrier Integrity, and Protective Actions* – satisfy the deterministic (structuralist[11]) [ACRS 1999] expectations for defense-in-depth. These protective strategies are the defense-in-depth safety fundamentals that complement the design objectives, as shown in Figure 5-1. This section describes each strategy and how they were chosen and why they form a sufficient set.

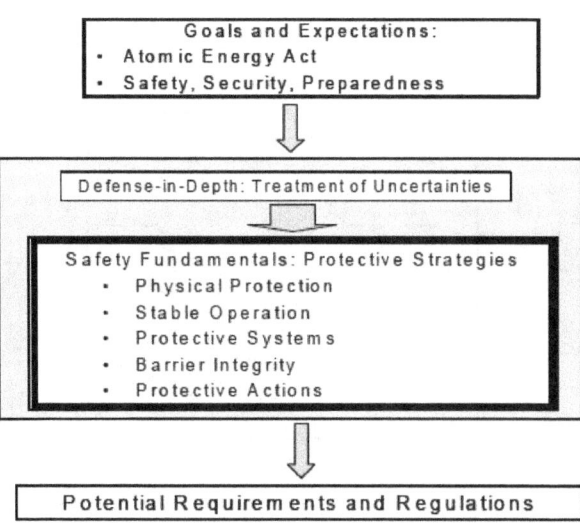

Figure 5-1   Safety Fundamentals:
Protective Strategies

The five protective strategies form a sufficient set for two reasons: they meet a set of minimal needs from an engineering perspective and they map to the physical pathways that need to occur in the plant, if damage is to occur.

The engineering perspective begins with the traditional view of defense-in-depth, the idea of multiple barriers to release: for light water reactors (LWRs) there was fuel bound in an oxide matrix,

---

[11]Defense-in-depth describes the deterministic approach as embodied in the structure of the regulations and in the design of the facilities.

clad in a tough alloy with good heat transport properties, contained within a leak tight primary coolant system, located inside a low leakage containment structure. If the design can maintain integrity of just one of these barriers, only very small amounts of hazardous materials are released. To protect the barriers, the design needs to have the capability to prevent damage and that can be ensured by maintaining stable operations (minimizing intrinsic challenges) and providing physical protection (to reduce the chance of successful extrinsic attack). If stable operations should be disturbed, protective systems and protective actions can terminate potentially dangerous event sequences. Finally, should the barriers be breached, protective systems and protective actions can mitigate the damage and minimize release.

Thus the five protective strategies provide layers of protection at all levels of engineering consideration.

Figure 5-2 shows another direct way to consider the protective strategies: the pathways to damage.

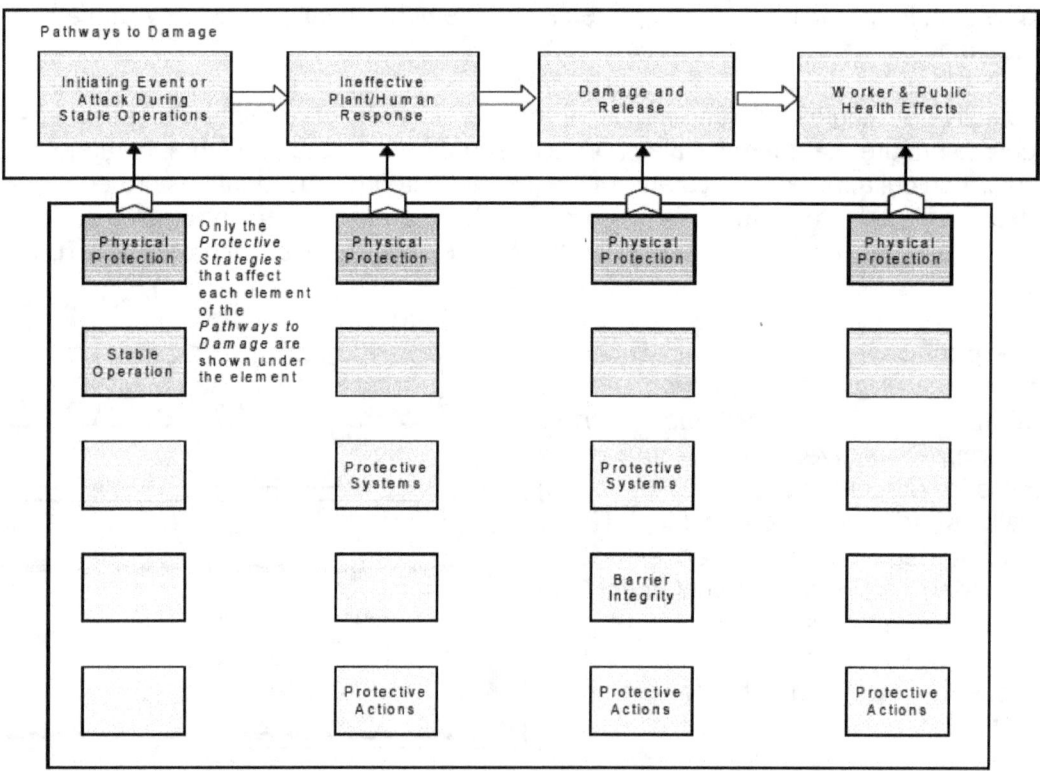

Figure 5-2   The Complete Nature of the Protective Strategies

The top chart shows that to reach a damage state, the plant needs to follow a physical pathway to damage, departing stable operations via an inadvertent initiating event or an attack. Any pathway to damage needs to involve the failure of plant equipment and operators to terminate the pathway before damage. Next, plant systems and operators need to arrest the release and propagation of radionuclides. Finally, the pathway needs to carry sufficient material to the location of workers and the public to cause health effects (injuries or fatalities). Note that at least two protective strategies can interfere with every stage of the pathway. By their interaction with all stages of pathways to damage, the five protective strategies are clearly a sufficient set.

An alternative way to view the physical pathways is to overlay the probabilistic risk assessment (PRA). Its purpose is to predict those physical pathways to damage that can occur. For every source of radioactive hazard on-site, the response to each possible initiating event is modeled in the PRA. Thus the PRA examines the ways in which multiple barriers[12] can be breached; it models:

- initiating events

- successes and failures in the protection systems that are designed to protect barriers

- human actions that can perform or defeat the protective systems or barriers themselves

- the physical response of the integrated plant to event sequences, including radiological dispersion pathways

- the emergency response system developed to protect the public and workers in case barriers fail

- dose response, calculating the probability of frequency of human health effects and land contamination

Each protective strategy interacts with one or more elements of the PRA model. PRA models of the protective strategies are based on technology-specific design and implementation, which is itself guided by the technical and administrative regulations that apply to design, construction, and operation. If the results of the PRA compare favorably with the safety/risk objectives, the protective strategies are adequate for the new technology system. Protective strategies add a layer of protection beyond that implied by the PRA. Because they are all required, they provide a high level defense-in-depth structure for identifying potential safety requirements, as described in Chapter 8. Furthermore, this layer of defense-in-depth provides a measure of protection against uncertainties, even those that are due to technical knowledge gaps that are not known and not modeled in the PRA.

The link between the protective strategies and actual regulation is established through an examination of the elements that affect each strategy. These are discussed below.

5.2   Analysis to Identify Potential Requirements

The process for identifying potential requirements is described in Chapter 8. It begins with identifying the protective strategies and focuses on ensuring that they are maintained throughout the life of the plant through efforts in design, construction, operations, and regulation. Figure 5-3 shows the process.

---

[12]Barriers include physical barriers and the physical and chemical form of the material that can inhibit its transport if physical barriers are breached.

If the protective strategies defined in this chapter are maintained, only small amounts   of radioactive material can be released. Therefore the next step is to identify those challenges or threats that could damage or disable one or more protective strategies. In the final step, potential requirements are developed for design, construction, and operations that will ensure integrity of the protective strategies. Chapter 8 describes this process and it is also outlined below.

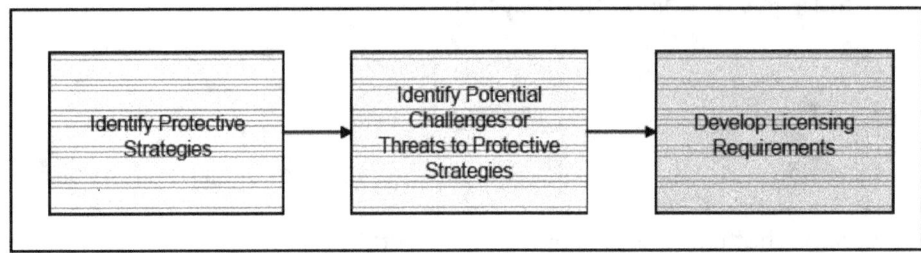

Figure 5-3   Process for Developing Potential requirements

Potential challenges to the five protective strategies are analyzed in Chapter 8. The approach is to develop a logic tree for each strategy, asking, how can this strategy (e.g., the set of barriers) fail to provide its function. This is a top-down analysis that begins by partitioning the functional failure into two or more classes of failure. It usually proceeds by identifying specific causes of failure. The basic structure of these logic trees is shown in Figure 5-4, where functional failure of a protective strategy is deductively examined by looking for failure to perform as required, failures through improper analysis or implementation, and unanticipated failures.

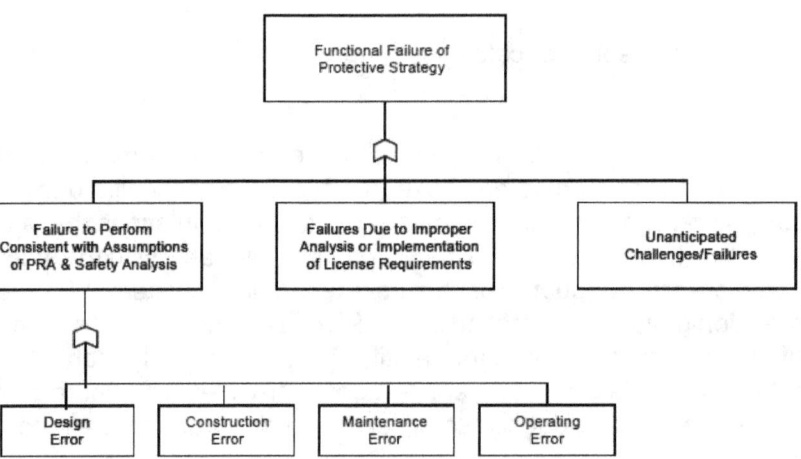

Figure 5-4   Logic Tree Developing Requirements for Each Protective Strategy

For each protective strategy, the logic tree is developed down to the level of specific failures to perform for the first branch; e.g., specific failures such as a design error to address corrosion or an operator error, because the procedures failed to account for all possible environments.

Next, these failure causes (bottom events in the logic tree) are examined for their relevance during design, construction, and operations. Questions are developed for regulators that, when answered, will identify the topics that will need to be addressed in the requirements for design, construction and operation, if the protective strategies are to remain functional. In developing the potential requirements themselves, a performance-based approach is used wherever practical. Details of this process are carried out in Appendices G and J, where potential technical and administrative requirements are developed to provide high assurance that the protective strategies can fulfill their functions.

5.3 The Protective Strategies

5.3.1 Physical Protection

The physical protection strategy provides measures to protect workers and the public against intentional acts (e.g., attack, sabotage, and theft) that could compromise the safety of the plant or lead to radiological release. Physical protection is provided by inherent design features and by extrinsic measures ("guns, guards, and gates") to provide defense-in-depth against attack. This requires that the design makes it unlikely that outsiders or insiders can reach sufficient sensitive areas of the plant to accomplish their goals, either using standoff weapons or by actual entry into the plant. Furthermore, the extrinsic features provide for detection, delay and response. Physical protection requires an integrated view of the plant and threats considered.

Physical protection can be tested by analysis similar to the safety PRA. [Garcia 2002] In this case one needs to characterize the threat, that is the type of threats and their objectives, capabilities, and strategies. While the likelihood of an attack is difficult to assess, it is possible to characterize the range of possible threats. Next the system, including its operators and safeguards, is evaluated against the range of threats to understand the possible scenarios.

A goal of the risk-informed, performance-based framework is to build physical protection into the plant design early in the design process to improve the intrinsic resistance of the plant and minimize the reliance on human actions.

5.3.2 Stable Operation

The stable operation strategy provides design and operations measures to make it unlikely that challenges to safety develop during operations. A thorough examination of potential initiating events is conducted as part of the risk analysis of the design. The initiating events are identified, along with their mean frequency of occurrence. Uncertainty in their frequency is also considered and quantified as a probability of frequency distribution. Initiating events will include events from both plant internal and external causes, as well as events during all operating states, since these are all in the scope of the risk analyses. Events that could affect any sources of radioactivity are modeled.

Initiating events vary in their potential impact. For example, an initiating event that simply trips an operating reactor is fairly benign, while common cause initiating events (those that directly challenge barriers or disable or degrade protective systems) require fewer additional failures before radionuclide release.

It may also be advantageous to group the initiating events into certain classes depending on their frequency of occurrence, as frequent, infrequent or rare. Such a grouping allows the protective features (addressed in the next protective strategy) to have reliability and performance that is commensurate with the frequency of the initiating events group, so as to limit the frequency of fuel damage accidents to acceptable levels.

For the future reactor technologies, initiating events may be substantially different from those for current U.S. LWRs. As described in Chapter 7, techniques can ensure that the search for initiating events is thorough and well structured.

5.3.3 Protective Systems

The protective systems strategy provides highly reliable equipment to protect plant functions, maintain barriers, and mitigate the effects of accidents and attacks. Plant features are provided to mitigate the consequences of initiating events by protecting the barriers identified in the following protective strategy. A critical part of the determination of these features is a qualitative review of the reactor-specific design philosophy, which includes a review of the design and performance features of the barriers, the reactor-specific safety functions that protect these barriers, the specific inherent and engineered safety features of the reactor concept in light of their capability to protect the barriers. Another critical part of the determination is the full scope (internal and external events, all operating modes) PRAs that are performed for the new designs. These PRAs not only determine the needed features, but also their required reliability and capability. The PRAs are used to demonstrate that the safety/risk objectives are within the desirable range, with adequate consideration of uncertainty.

For some scenarios which appear to be credible but have very broad uncertainty (due to insufficient data, not well understood phenomena,) additional protective features may need to be incorporated. If licensing basis events (LBEs) are needed to address such scenarios, as described in Chapter 6, then the protective features necessary to cope with the LBEs are identified and incorporated.

For the future reactor technologies, some mitigative considerations are substantially different from those for current US LWRs and can be appropriately modeled, as described in Chapter 7.

5.3.4 Barrier Integrity

The barrier integrity strategy provides isolation features that protect the primary radionuclide inventory from release. Functional barriers to radionuclide release are provided to maintain isolation of hazardous nuclear material within the system. Barriers can be both physical barriers and barriers to mobilization and transport of radioactive material, e.g., the physical and chemical form that retards the dispersion of the material. Again, the plant PRA can play a critical part in determining the number and type of these barriers, as well as their required reliability and capability. The PRAs are used to demonstrate that the frequency of radionuclide release is low enough, with adequate consideration of uncertainty. Uncertainties associated with barrier degradation, e.g., corrosion, erosion, aging, and other materials issues, need to be modeled. For some systems, chemical interactions are important.

Additional barriers, beside those identified from the risk analysis, may be needed to address credible scenarios not amenable to risk analysis and identified as LBEs (Chapter 6). They may be needed to provide assurance against uncertainties in modeling completeness as well.

5.3.5 Protective Actions

The protective actions strategy provides planned activities that protect the other strategies and, should those strategies fail in spite of attempts to protect them, mitigate the impacts of their failure. "Preparedness" is a function of how well procedures are written, how well personnel are trained, and how accessible needed equipment and personnel are. Protective actions are in place to protect the public, even if all design features fail and a release of radionuclides from the plant occurs.

Should functional barriers fail to adequately limit the radionuclide release, protective actions are provided to control the accident progression, and ultimately, to limit the public health effects of accidents. The analysis of the plant PRA helps to determine the measures that are effective in limiting the public health effects from radionuclide release accidents so that the risk remains below the quantitative health objectives.

Protective actions include actions of operators in response to departures from stable operations (i.e., actions specified in abnormal and emergency operating procedures), actions by personnel in the emergency response center (as prescribed by the Severe Accident Management Guidelines), actions by the security team in response to an attack, on-site health physics management of radiological hazards to workers, and the actions of first responders, state, and local officials in accordance with emergency plans.

5.4   References

[ACRS 1999] Letter D. Powers to S. Jackson, "The Role of Defense in Depth in a Risk-Informed Regulatory System," ACRS, May 19, 1999.

[Garcia 2002] Garcia, M.L., *The Design and Evaluation of Physical Protection Systems*, Butterworth-Heinemann, Boston, 2002.

6.   LICENSING BASIS: PROBABILISTIC PROCESS

6.1   Introduction

The purpose of this chapter is to outline the probabilistic process included in the licensing approach that is part of the risk-derived, performance-based framework, applicable to all reactor designs. Figure 6-1 shows the place of this chapter in the Framework document structure.

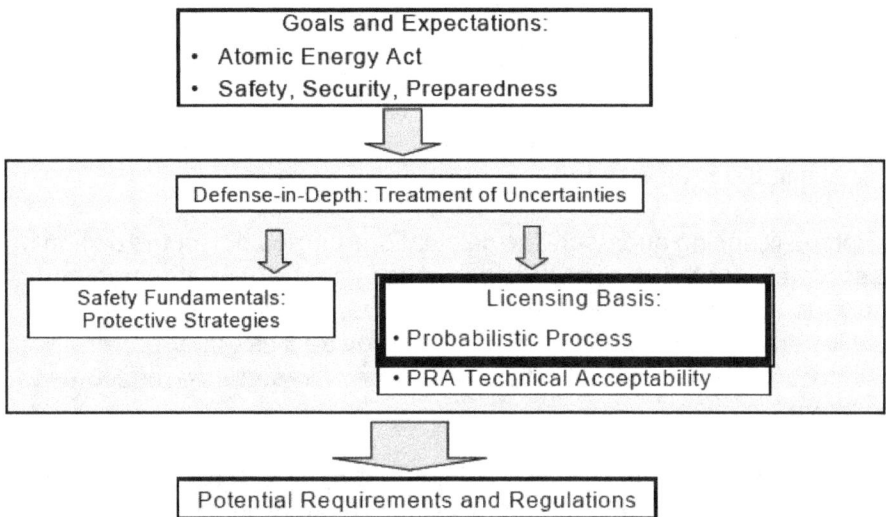

Figure 6-1   Framework Licensing Basis: Probabilistic Process

The process described below has several objectives to:

*   Specify the acceptability of the estimated plant risk (Section 6.2),

*   Develop a frequency-consequence (F-C) curve to be used in selecting Licensing Basis Events (LBEs), and in establishing criteria for the LBEs to meet. (Section 6.3),

*   Develop a procedure, i.e., criteria and guidance, for identifying and selecting a complete set of LBEs (Section 6.4),

*   Develop a procedure for classifying risk-significant systems, structures and components (SSCs) to ensure that the reliability and functionality of the SSCs are consistent with their design, and their intended maintenance and operation (Section 6.5),

*   Ensure that adequate regulatory margin exists and to encourage the use of additional operational margins (Section 6.6), and

*   Integrate security into the design process at least at the same level of protection as established by the post 9/11 U.S. Nuclear Regulatory Commission (NRC) requirements (Section 6.7).

The process developed below is meant to provide guidance to the NRC staff to judge the acceptability of a licensing application.

To focus the discussion in the rest of this chapter, it is useful to ask: "What are the differences, at a high level, between the approach in this document and the current approach to licensing as set out in Parts 50 and 52 of Title 10 of the Code of Federal Regulations (10 CFR)?"

For both the current approach and the approach in this document, an applicant needs to submit a Safety Analysis Report (SAR) to construct or operate a nuclear power plant. The SAR contains the design criteria and information for the proposed reactor plant and comprehensive data on the proposed site. The SAR also identifies and analyzes various hypothetical accidents and the safety features of the plant which prevent accidents or, if they should occur, mitigate their effects. The selection process of these hypothetical accidents, and of the systems, structures, and components (SSC) that prevent or mitigate these accidents, i.e., are important to plant safety, are carried out via a new approach in the Framework.

In the current Part 52 licensing approach, the calculations carried out for the design basis accidents (DBAs) and, separately, for the probabilistic risk assessment (PRA) are important components of the safety analyses, but there is no direct link between these two components. (The recent development of [voluntary] Part 50.69 however, does provide a risk-informed link between the PRA and the SSC selection for safety classification by the use of importance measures to identify SSCs for special treatment.)

The approach in this document is a more risk-informed approach since it links the PRA analysis with the other design objectives of licensing basis event selection and criteria, and selection and treatment of SSCs. This NUREG uses an approach with a probabilistic component, i.e., the information from the PRA analysis, for

(1)   Selecting the licensing basis events (this document uses the term 'licensing basis events [LBEs]' instead of DBAs since the LBEs include risk-significant events in the licensing basis down to a mean frequency of 1×10^{-7} per year, which have different acceptance criteria, dependent on their frequency, as explained further below), and

(2)   Selecting systems, structures, and components for which special treatment is needed because of their safety significance (i.e., in maintaining risk below the chosen acceptance criteria).

(3)   Replacing the traditional "single-failure criterion."

This document still relies on deterministic and defense-in-depth considerations in both the LBE selection and criteria, and the SSC selection and treatment, but, in addition, uses the risk information from the PRA to focus attention on the risk-significant aspects of the design.

The methodology and criteria for implementing the probabilistic approach to licensing is considered a policy issue that needs the Commission's review and direction. Appendix C discusses this issue further.

6.2   Acceptability of Plant Risk

In the approach in this document, a probabilistic risk assessment (PRA) is used as part of the licensing of plants to generate a sufficiently complete set of accident scenarios whose frequencies and consequences, individually and cumulatively, provide an estimate of the overall risk profile of the plant. The question is what constitutes an acceptable plant risk as estimated with the PRA?

The scope of a PRA used in this document is broader than the scope of PRAs that have been conducted for the current generation of plants. Since the safety and licensing basis for the current fleet of plants was established without the benefit of PRA, and due to the nature of the Light Water Reactor (LWR) risk profile, the scope of PRAs for LWRs has been mostly confined to the analysis of beyond design basis core damage accidents. For the PRA of a future reactor, the Framework approach expects that the PRA will provide important input to selecting licensing basis events (see Section 6.4), so the scope is broader and is not necessarily limited to very low frequency event sequences.

In addition, although risk can be generally expressed in terms of consequences resulting from exposures, in LWR PRAs the risk is often expressed in terms of surrogate measures such as core damage frequency (CDF) or large early release frequency (LERF). These surrogate measures and the criteria associated with them (such as CDF < 1E-4/yr or LERF <1E-5/yr for some applications) are LWR-specific and not applicable to all reactor designs. Finally, since frequent, infrequent, as well as rare events are included in the PRA, a single limiting criterion on consequence or its surrogate (such as CDF or LERF for LWRs) may not be adequate. Instead, a criterion that specifies limiting frequencies for a spectrum of consequences, from none to very severe, needs to be established. A frequency-consequence (F-C) curve, that relates allowable consequences to their allowable frequencies, provides such a criterion. Such a curve is proposed in the next section as a criterion for selecting licensing basis events and judging their acceptability. Besides the acceptability of the individual licensing basis events, the total or integrated plant risk needs to also be considered and found acceptable. One criterion for an acceptable integrated risk is compliance with the NRC safety goals and their associated Quantitative Health Objectives (QHOs).

6.2.1   Compliance with the Quantitative Health Objectives

The following are definitions of the Quantitative Health Objectives (QHOs) taken directly from the Safety Goal Policy Statement [NRC 1986]:

- "The risk to an average individual in the vicinity of a nuclear power plant of prompt fatalities that might result from reactor accidents should not exceed one-tenth of one percent (0.1%) of the sum of prompt fatality risks resulting from other accidents to which members of the U.S. population are generally exposed."

- "The risk to the population in the area of nuclear power plant of cancer fatalities that might result from nuclear power plant operation should not exceed one-tenth of one percent (0.1%) of the sum of cancer fatality risks resulting from all other causes."

The average individual risk of prompt (or early) fatality and latent cancer fatality that is calculated in the PRA to compare with the safety goals and the QHOs is the total plant risk incurred over a reactor year. This means the PRA results need to demonstrate that the total plant risk, i.e., the risk summed over all of the accident sequences in the PRA, need to satisfy both the latent cancer QHO and the early fatality QHO. The safety goals, and consequently, the QHOs are phrased in terms of the risk to an 'average' individual in the vicinity of (or 'area near') a nuclear power plant per reactor year. The latent cancer QHO is defined in terms of the risk to an average individual within 10 miles and the early fatality QHO in terms of the risk to an average individual within 1 mile of the plant.

Therefore, the PRA results need to show that the total integrated risk from the PRA sequences satisfy both the latent cancer QHO and the early fatality QHO.

One question raised about the safety goals is whether they apply to a single unit or to the whole site, i.e., if there are multiple units at a site, then should the integral risk from multiple units meet the safety goals? As the statement above indicates, the QHOs address the risk to an individual that lives in the 'vicinity' of a nuclear power plant. If the plant consists of multiple units, then the individual is exposed to the risk from those units, and therefore, the site. The approach of this document is that the integrated risk posed by all new reactors at a single site should not exceed the risk expressed by the QHOs. This is complementary to the minimum level of safety recommended for new reactors in Chapter 3. Both the individual risk of each new reactor and the integrated risk from all of the new reactors at one site, associated with a future combined license application, should not exceed the risk expressed by the QHOs. The question of integrated risk is considered to be a policy issue that the Commission needs to review and s discussed further in Appendix C.

This approach does not require that the integrated risk from existing reactors, where there are multiple reactors at a single site, meet the risk expressed by the QHOs, even though the site may be considered for new reactors. In the Policy Statement on Severe Reactor Accidents Regarding Future Designs and Existing Plants [NRC 1994]," the Commission concludes that existing plants pose no undue risk to public health and safety and sees no present basis for immediate action on generic rulemaking or other regulatory changes for these plants ...." This statement is supported by the Commission's policy statement on Safety Goals for the Operation of Nuclear Power Plants that states that current regulatory practices are believed to ensure that the basic statutory requirement, adequate protection of the public, is met. In considering new plants at a site with or without existing plants, it should be assured that the new plants pose no undue risk to the public. Limiting the integrated risk for new plants to the risk expressed by the QHOs (and thereby imposing enhanced safety for these new plants), ensures that the new plants pose no undue risk to the public.

6.2.2 Other Possible Measures of Integrated Risk

In this document in Chapter 3, compliance with the NRC safety goals is mandatory. Therefore, in the approach of this NUREG the total plant risk is acceptable if the Quantitative Health Objectives (QHOs), as described in the NRC's Safety Goal Policy Statement, are met. The QHOs provide a limiting value annually of early fatalities and latent cancer fatalities. They are an integral measure of risk over the entire frequency spectrum of event scenarios. In addition to the QHOs, further requirements could be specified for the total risk profile, for example, in terms of an acceptable complementary cumulative distribution function (CCDF) for a risk measure of interest, e.g., early fatalities. Such a CCDF would provide a regulatory limit for the appropriate risk measure as a function of frequency. Possible benefits from such an acceptable curve as criterion, and the plant-specific CCDF calculated to meet the criterion, could include:

(1)   complementing the F-C curve discussed in Section 6.3.2 by ensuring that the cumulative risk from high-frequency events is low (whereas the F-C curve ensures that the consequences from individual high-frequency events are low),

(2)   Providing insight into the design-specific distribution of risk,

(3)   Providing the basis for quantitatively establishing the desired relation between accident prevention and mitigation, and

(4)   Providing a criterion for assessing the integrated effect of safety, security, and preparedness.

Currently, no CCDF or other requirement besides the QHOs is imposed on total risk by this document's approach. Such requirements may be developed in the future in terms of health effects or of surrogate risk measures for particular technologies. Appendix C discusses this issue of the use of a CCDF curve further.

While it does not currently provide additional requirements on the total risk distribution, the approach of this document does provide a frequency versus consequence (in terms of dose) curve that sets dose and frequency limits for individual sequences or groups of sequences. This curve, which is a tool for LBE selection and LBE criteria, is derived on the basis of existing dose limits found in the regulations.

6.3   Frequency - Consequence Curve

A criterion that specifies limiting frequencies for a spectrum of consequences, from none to very severe, can be denoted via a frequency consequence (F-C) curve. On the F-C plane, the F-C curve provides an acceptable limit in terms of the frequency of potential accidents and their associated consequences. The objective of the F-C curve is to establish the licensing basis, i.e., identify the event sequences that the design and operation of the plant need to be able to mitigate. This objective involves first establishing criteria for ensuring that the frequency of occurrences of event sequences is inversely related to the consequences, e.g., high-frequency events such as anticipated operational occurrences (AOOs),i.e, a turbine trip with successful mitigation, have low consequences, and high-consequence events like containment bypass have low frequency. Second, the objective involves establishing criteria that define the acceptable frequencies for different levels of consequences.

6.3.1   Considerations in the Development of a F-C Curve

The range of frequencies that need to be considered in establishing the F-C curve encompasses a wide range that includes frequent events (AOOs), infrequent events, and rare events. The consequences can be expressed in several different units of measure which include released activity in terms of curies (or becquerels) of various radionuclides, health effects like early fatalities and latent cancers, and radiation doses (rems or sieverts). Each of these options is briefly discussed below.

To develop an F-C curve with consequences defined by curies would require establishing acceptable limits of releases of various radionuclides because given amounts of curies associated with different radionuclides lead to different doses depending on the mode of exposure (cloudshine, groundshine, inhalation, ingestion.), and hence different health effects. One can attempt to convert the released curies of different radionuclides into *equivalent* curies of one particular radionuclide (e.g., I-131) but this introduces additional complexity. The notion of equivalent curies is essentially based on the concept of (effective) dose equivalence; since radionuclides are released into the atmosphere as either gases or aerosols, their transport and dilution off-site depends on the site's meteorological characteristics (windspeed, stability class, rainfall, etc.) so that what would constitute equivalent curies at one site would not necessarily be the same at another site. One can avoid this variability by defining a generic site from a meteorological standpoint, but this definition would need to gain acceptance since it may either penalize an actual site proposed by a licensee (if the site has "better" meteorology) or allow a weakness in the design (if the actual site is "worse"). In any case, a site-specific consequence assessment (i.e., a Level-3 type PRA) would still be needed to confirm that the results based on the generic site are met by the actual site. Apart from complexity, the problem with establishing consequence limits in terms of curie releases is that, unlike radiation doses, there are no pre-established limits to draw upon. Hence acceptable curie limits may be difficult to understand, communicate, or defend.

Another possible unit of measure of consequences in developing the F-C curve is health effects, e.g., early fatalities and latent cancer fatalities. The risk of these health effects was used by the Commission in its Safety Goal Policy Statement of 1986 to establish acceptable risk goals for plant operation, however, they were measured in terms of the quantitative health objectives (QHOs). The QHOs are point values and express the risk to an average individual of either early fatality (within 1 mile of the plant boundary) or latent cancer fatality (within 10 miles of the plant boundary) where the average individual risk is calculated by estimating the (conditional) number of early/latent fatalities to the appropriate distance divided by the total population to that distance. To establish an F-C curve, where consequences are measured in absolute numbers of health effects, would require a definition of societal risk, i.e., the establishment of an acceptable limit for the absolute number of health effects (early fatalities or latent cancer fatalities) as a function of frequency at a particular distance from the plant boundary. (This number cannot be derived from the QHOs which express the average individual risk of health effects summed over all event frequencies). Commission policy has been to derive limits for average individual risk of health effects based on comparisons to societal concepts that are easily understood and communicated, e.g., one-tenth of one percent of the annual rate of cancer fatalities in the U.S. for deriving the latent cancer fatality QHO. Establishing societal health effects, however, in terms of an acceptable limit on the absolute number of early or latent fatalities is inconsistent with Commission policy. In any case, if such limits are established, the licensee will need to demonstrate compliance at an actual site or a generic site and, if the latter is used, a Level-3 type PRA would still be needed to confirm that the results based on the generic site are met by the actual site.

Radiation dose (rems, sieverts) is yet another measure of consequence to use in developing the F-C curve. One advantage of this measure is that it is based on national and international regulatory practice, e.g., NRC regulations in 10 CFR 20 and 10 CFR 50, EPA (Environmental Protection Agency) protective action guidelines, IAEA guidelines and International Commission on Radiation Protection (ICRP) recommendations. However, the guidance that exists in terms of actual limits and how they are expressed is occasionally inconsistent since different values were developed at various times to serve different objectives. Despite this shortcoming, the main advantage of using dose for consequence is that a body of rules and guidance already exists for its use while the use of other measures could result in confusion. Like other consequence measures, to demonstrate conformance to regulatory limits the licensee will need to use an actual site and, if a generic site is used, a Level-3 type PRA would still be needed to confirm that the results based on the generic site are met by the actual site.

For all the measures of consequence discussed above, the F-C curve only addresses the consequence limits for individual sequences or groups of sequences, it does not provide a total risk distribution.

6.3.2   A Proposed F-C Curve

Based on the considerations discussed above, the consequences of the F-C curve adopted for this document are expressed in terms of radiation dose at the plant site boundary, since dose is an example of a metric that can be directly linked to consequences.

An F-C curve, that is based on, and derived from, current regulatory requirements in Parts 20, 50, and 100, was developed. While these existing requirements provide some dose anchor points, considerable engineering judgement is used to assign corresponding frequencies to establish the F-C curve. Therefore, at the outset other F-C curves could be developed and justified based on engineering judgement.

Part 20 limits the radiation doses from licensed operation to individual members of the public. Part 50 Appendix I identifies design objectives for releases during normal operation to be as low as reasonably achievable (ALARA). Both of these regulations are concerned with cumulative dose acquired annually, rather than during a single event. Part 50.34 requires an applicant for a license for a power reactor to demonstrate that doses at the site boundary (and the outer boundary of the low population zone) from hypothetical accidents (i.e., per event) will meet specified criteria. Part 100 has similar dose criteria for determining site suitability.

Based on these existing requirements, two F-C curves could be developed: (1) one in the high-frequency-low consequence range based on cumulative (annual) requirements, which would show rems per year on the ordinate and probability on the abscissa; (2) a curve in the lower-frequency-higher consequence range based on dose limits for individual accidents. In this document a single, per event, F-C curve is used and the cumulative requirements at high frequency are imposed as added limitations to the per event curve.

The principle underlying the F-C curve is that event frequency and dose are inversely related, i.e., the higher the dose the lower is the event frequency. This principle, and the whole F-C curve, is broadly consistent with the approach of ICRP 64. Recommendations on the annual frequencies and doses to individual members of the public from accidental exposures are provided in ICRP 64. The doses cover a wide range of severity, from small exposures that are within regulatory limits to very high exposures that can lead to an early fatality [ICRP 1993].

10 CFR 50 Appendix I provides numerical guidance for doses that are ALARA from normal operation of nuclear power plants. The recommended value is 5 mrem per year whole body (or, equivalently, 5 mrem per year total effective dose equivalent [TEDE]) to an individual in an unrestricted area, thus in the F-C curve used here doses in the range of 1 mrem - 5 mrem are assigned a frequency of 1 per year.

10 CFR 20 limits public exposure from licensed operation to 100 mrem in any one year and the range from 5 mrem - 100 mrem is assigned a frequency of 1E-2 per year on the F-C curve (events in this category would generally constitute what are currently known as anticipated operational occurrences or AOOs).

The next higher dose category ranges from 100 mrem to about 20-25 rem. This category involves doses that are above public limits for licensed operation but only involve stochastic health effects. Doses in the range of 100 mrem to 25 rem are subdivided into two ranges: those below the EPA protective action guideline [EPA 1992] of 1 rem off-site are assigned a frequency of 1E-3/year while those in the range of 1 rem to 25 rem are assigned a frequency of 1E-4 per year. The DBA off-site dose guideline in 10 CFR 50.34 and 10 CFR 100 is 25 rem; it is also the dose that defines an abnormal occurrence (AO) as described in the Commission's April 17, 1997, policy statement on AOs, (62 FR 18820 [NRC 1997]) which defines substantial radiation levels to imply a whole-body dose of 25 rem to one or more persons. (As part of the safety analysis, 10 CFR 50.34 requires the applicant to assume a fission product release from the core and determine that an individual located on the boundary of the exclusion area for a period of 2 hours following the onset of the release, or on the outer boundary of the low population zone for the entire duration of the passage of the plume resulting from the release, would not receive a radiation dose in excess of 25 rem total effective dose equivalent [TEDE]).

Doses above 50 rem fall in a category where some radiation effects are deterministic (ICRP 41). [ICRP 1984] gives a threshold of 0.5 Sv, 50 rem, based on 1% of the exposed population showing the effect for depression of the blood-forming process in the bone marrow, from whole-body exposure). Thus, doses in the range of 25 rem - 100 rem are assigned a frequency of 1E-5 per year. Doses where early fatality is possible are characterized by a threshold (e.g., lethal dose to 1% of the population) and an LD_{50} value (median lethal dose). For bone marrow syndrome from whole-body exposure, the threshold dose is 1 Sv (100 rem), for a population receiving no medical care and 2-3 Sv (200 - 300 rem) for a population receiving good medical care. In the NRC-sponsored MACCS probabilistic consequence analysis code [NRC 1998], the threshold and LD_{50} parameters for early fatality due to bone marrow syndrome are set at 150 rem and 380 rem, respectively for a mixed population consisting of 50% receiving supportive medical care and 50% receiving no medical care, based on the early health effects models developed in NUREG/CR-4214 [NRC 1989]. Based on these considerations, doses in the range 100 rem - 300 rem are assigned a frequency of 1E-6 per year, 300 rem - 500 rem a frequency of 5E-7 per year, and the curve is capped beyond doses greater than 500 rem at 1E-7 per year.

These values are shown below in Table 6-1 and are plotted in Figure 6-2.

Table 6-1   Proposed dose/frequency ranges for public

Dose Range	Frequency (per year)	Comment (all doses are TEDE)
1 mrem - 5 mrem	1E+0	5 mrem/year is ALARA dose in 10 CFR 50 App I
5 mrem - 100 mrem	1E-2	100 mrem/year is the public dose limit from licensed operation in 10 CFR 20
100 mrem - 1 rem (1)	1E-3	1 rem/event off-site triggers EPA Protective Action Guidelines (PAGs)
1 rem - 25 rem (1)	1E-4	25 rem/event triggers AO reporting and is limit in 10 CFR 50.34a and in 10 CFR 100 for siting
25 rem - 100 rem	1E-5	50 rem is a trigger for deterministic effects (i.e., some early health effects are possible)
100 rem - 300 rem	1E-6	In this range the threshold for early fatality is exceeded
300 rem- 500 rem	5E-7	Above 300 - 400 rem, early fatality is quite likely
> 500 rem	1E-7	Above 500 rem early fatality is very likely and curve is capped

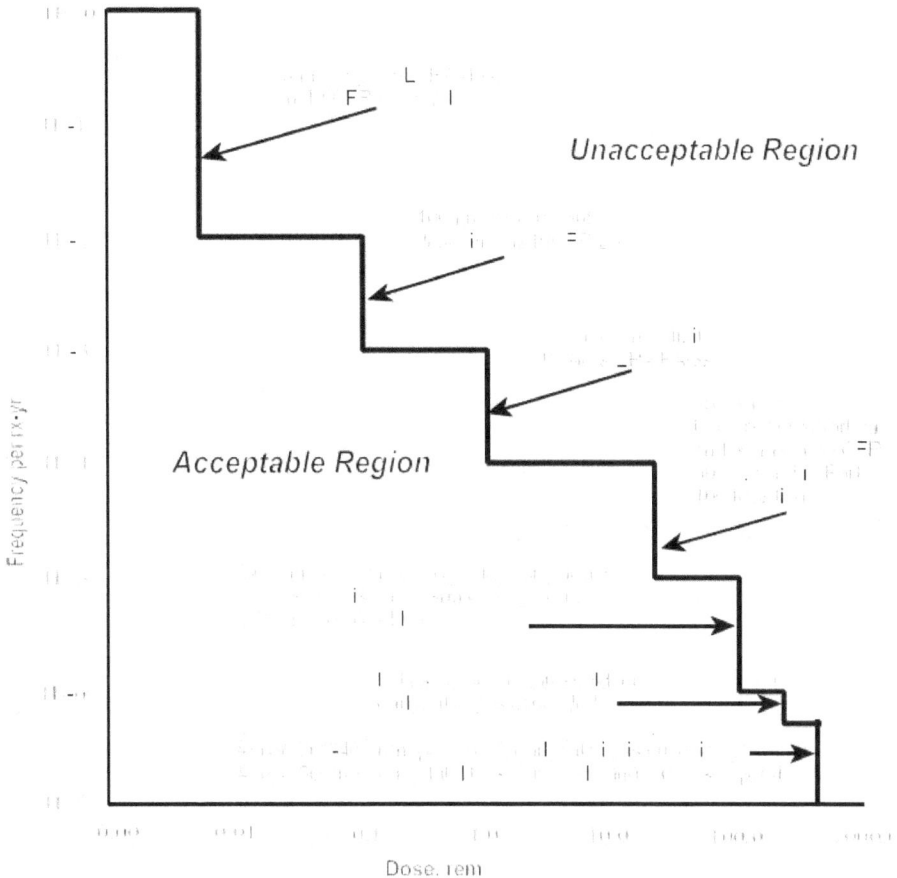

Figure 6-2 Frequency consequence curve

As previously noted, Appendix I of 10 CFR Part 50 and 10 CFR Part 20 are used above to construct the dose limits at the higher frequencies, i.e., frequencies greater than 1E-3 per year, of the F-C curve. The dose limits specified by Appendix I of Part 50 and by Part 20 are cumulative (over one year) dose limits for normal operation, while the other dose references used for the F-C curve are per event doses from accidental or unplanned exposures. Since the F-C curve is meant to provide criteria for unplanned exposures, i.e., exposures on a per event basis, the use of Appendix I and Part 20 may seem questionable for the F-C curve, in that it blurs the distinction between normal and unplanned exposures, but no per event criteria for unplanned exposures exist at low doses. The cumulative nature of the criteria in Appendix I and Part 20 will also play a role in the criteria imposed by this document on LBEs, as discussed in Section 6.4.

6.3.3   Meeting the Proposed F-C Curve

The sequences of the PRA will populate the space under the F-C curve. Some sequences will have little or no consequences, primarily because of the inherent characteristics and design features of the plant. Others are likely to approach the F-C curve and thus make up the important contributors to the plant risk profile. To be acceptable, the frequency and consequences of all the accident sequences examined need to lie in the acceptable region (i.e., below) of the F-C curve by meeting the dose criteria. As discussed later, this is the first step in selecting the licensing basis events.

Note that with the kind of acceptance criterion for individual sequences described above, an accident sequence is acceptable even though it has a dose at the boundary associated with it, as long as its dose and corresponding frequency do not exceed the limits specified by the F-C curve. For current LWR PRAs a single-limiting frequency of a surrogate risk metric, i.e., CDF or LERF, associated with potentially high-consequence event sequences, is an acceptance criterion. For the PRAs required by this document, whose scope covers all types of off-normal event sequences, the criterion is a series of limiting frequencies whose permitted value depends on the magnitude of their associated consequences. As illustrated in Figure 6-2, event sequences with high frequencies need to lead to no consequences or very minor ones; event sequences that are rather infrequent can have somewhat higher doses associated with them, and rare (very low frequency) event sequences can have higher consequences still. Meeting the F-C curve imposes additional constraints in addition to satisfying the QHOs because specific dose limits are imposed at all frequencies. For specific technologies it may be possible to eventually develop surrogate metrics (comparable to CDF and LERF for LWRs) for the dose parameter, along with acceptable values for such surrogates.

6.4   LBE Selection Process and LBE Criteria

The purpose of the LBEs is similar to the purpose of the DBAs [NRC 1996] in the current licensing process:

1.   to provide assurance that the design meets the design criteria for various accident challenges with adequate defense-in-depth (including safety margin) to account for uncertainties, and

2.   to evaluate the design from the standpoint of the dose guidelines in the siting criteria of 10 CFR Part 100.

This NUREG includes probabilistic selected LBEs that address the first requirement and a deterministic selected LBE that addresses the second requirement.

As described below, the PRA is used to select most of the LBEs, but the LBEs have to meet additional criteria, besides satisfying the F-C curve.

6.4.1   Probabilistic LBE Selection

The event sequences that make up the LBEs are selected from the PRA sequences. Before LBE selection, it is assumed that a complete PRA of the plant design covering both internal and external events and all modes of operation has been performed and that all accident sequences have been identified in terms of a distribution of their frequencies and end-states that are defined through consequences, estimated by the doses at the EAB.[13] The results have to meet the criteria of the proposed F-C curve, i.e., the frequencies and consequences of all sequences have to lie in the

---

[13]For the designer to make these calculations, they need to either be for a particular site or NRC needs to define a reference site, with sufficient detail to ensure it is adequate for any U.S. site. (i.e., consequences for the reference site will be greater than for real sites). An 80-th percentile site with respect to weather (i.e., consequences greater than at 80 percent of existing reactor sites with respect to variability of weather alone) is defined in NUREG/CR-6295 "Reassessment of Selected Factors Affecting the Siting of Nuclear Power Plants."

acceptable region of the F-C curve.[14] It is also assumed the PRA meets the appropriate review criteria, standards as outlined in Chapter 7.

The probabilistic selected LBEs provide a means to ensure that the design meets the design criteria for various accident challenges with margin. These sequences are derived from a design-specific PRA through a process described later in this chapter. The probabilistic selection process helps to ensure that the LBEs represent all potentially risk significant accident challenges.

The probabilistic selected LBEs are event sequences which represent challenges to plant safety. They encompass a whole spectrum of off-normal events including frequent, infrequent, and rare initiating events and event sequences. They include a spectrum of releases from minor to major, and sequences that address conditions less than the core damage sequences of the current reactors and those similar to current reactor core damage sequences.

In the case of LWRs, for example, this characterization would subsume events, similar to AOOs, involving either no release or very small amounts of release (e.g., from iodine spiking events, gap release.). It would also include design basis accidents as described in Part 50.34 or Part 100.11. (These are accidents where significant core damage is assumed to occur such that a large quantity of fission products is assumed to move from the fuel pellets/fuel rods to the reactor coolant system or the reactor vessel and ultimately into containment, but the containment is assumed to remain intact). Finally it includes large releases, i.e., accidents where a significant quantity of fission products is released from containment into the environment with a potential for causing early health effects to the off-site population.

As described later, the LBEs are sequences from the PRA that have to meet stringent acceptance criteria related to the F-C curve, and they need to also meet further deterministic criteria in addition to meeting the F-C criteria.

One of the stringent criteria attached to the LBEs is that SSCs credited in the LBE analysis will be considered safety-significant in this document.. All functions performed by the various plant SSCs that are included in the PRA have the potential to influence the frequency of LBE sequences and many influence the consequences. Therefore, any function and the associated SSCs included in the PRA used to arrive at successful end-states associated with the set of LBEs is safety-significant unless its failure probability has been set to 1.0 for guaranteed failure. The resulting PRA of the entire as-built plant needs to meet the F-C curve and the defense-in-depth deterministic requirements, as discussed later.

The designer decides as to what SSCs will be considered safety-significant as long as the Framework's acceptance criteria are met. This determination could be accomplished through an iterative approach, where the affect on the selection of LBEs is evaluated with a proposed set of safety-significant SSCs, then re-assessed with another set of safety significant SSCs, until the desired set of LBEs and other design objectives are achieved. Only safety-significant SSCs are credited for prevention or mitigation, but all systems whose failure or operation can initiate or

---

[14]There is a concern that having to demonstrate compliance with the F-C curve at very low dose levels via a PRA may unduly increase requirements for the scope and level of detail in the PRA performed to demonstrate compliance. To address this concern the PRA could be required to meet the F-C curve only below a frequency of 1E-3/yr, for example. This may be another reason for effectively breaking the F-C curve into two curves: one in the >1E-3/yr frequency range with cumulative dose criteria, and one for the rest of frequency range, below 1E-3/yr with per event criteria. A trial application of the Framework to an actual advanced reactor design would be useful to explore the practicality of these concepts.

aggravate an accident need to be included in the PRA model. Since the safety-significant SSCs are linked to the LBEs, and the LBEs were chosen in a risk-informed manner, this document's approach for selecting SSCs for special treatment is also risk-informed.

The selection of LBEs based on event sequences from the PRA also serves as a more realistic assessment of the impact on risk of equipment failures that are modeled in the analysis. As such, it serves as a replacement for the traditional "single failure criterion" applied in the current licensing process. Accordingly, in the criteria that are proposed in the Framework, a single-failure type of criterion is included only for defense-in-depth purposes for key safety functions, such as redundant, diverse, independent means of reactor shutdown.

In summary, the goal of the probabilistic selection process is to identify a set of bounding event sequences (i.e., sequences that essentially bound all the PRA sequences in terms of frequency and consequence) that demonstrate adequate defense-in-depth and safety margin from the standpoint of public health and safety. Figure 6-3 shows this process with each step described below.

Step 1 **Modify the PRA to only credit those mitigating functions that are to be considered safety significant.**

As stated above, those SSCs whose functionality plays a role in meeting the acceptance criteria imposed on the LBEs define the set of safety-significant SSCs. The SSCs of interest are those that reduce the frequency or limit the consequence of the LBEs, or both. All preventive or mitigative functions included in the PRA have the potential to influence the frequency of LBE sequences, and many influence the consequences. Therefore, any function and the associated SSCs included in the PRA used to develop the set of LBEs is safety-significant unless it has been set to 1.0 or guaranteed failure. The designer can remove mitigation functions from the PRA to reduce the set of safety-significant SSCs. In this way the designer can decide, most likely through an iterative process, which SSCs will be the safety-significant ones to meet the acceptance criteria. However, all systems whose failure or operation can initiate or aggravate an accident need to be modeled in the PRA. The resulting PRA needs to meet the F-C curve and the defense-in-depth deterministic requirements.

Step 2 **Determine the point estimate frequency for each resulting event sequence from the quantification of the modified PRA.**

Drop all PRA sequences with point estimate frequency < 1.E-8/yr. This step establishes the complete set of event sequences that will be processed to determine the LBEs.

Step 3 **For sequences with point estimate frequencies equal to or greater than 1E-8, determine the mean and 95th percentile frequency.**

The frequency used to determine whether an event sequence remains within scope of the LBE selection process is based on the 95th percentile. Therefore, the mean and 95th percentile are determined in this step.

Step 4 **Identify all PRA event sequences with a 95th percentile frequency > 1E-7 per year.**

This step identifies those sequences that are to be included in the event class grouping process. Sequences less than 1E-7 per year are screened from the process.

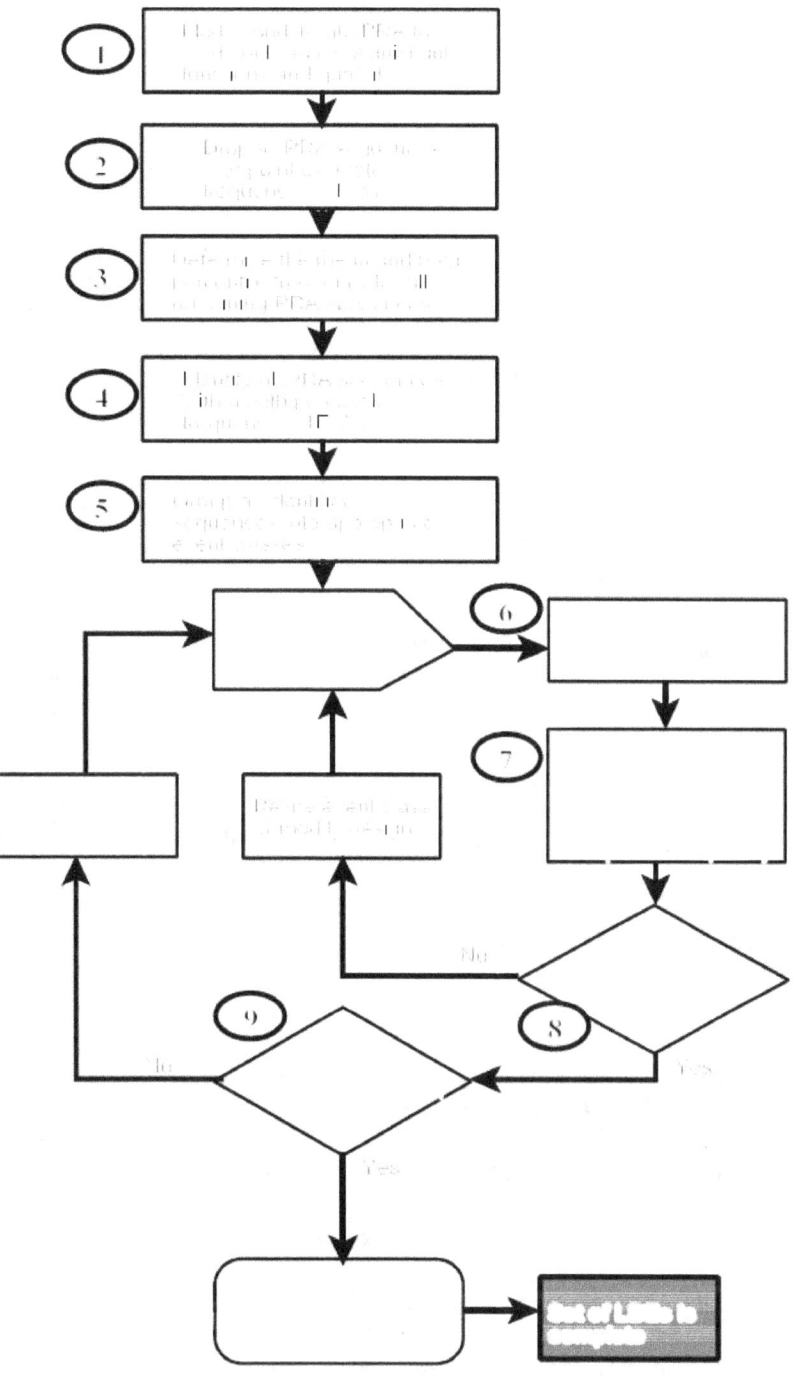

Figure 6-3   Licensing Basis Event Selection Process

Step 5 **Group the PRA event sequences with a 95th percentile frequency > 1E-7 per year into event classes.**

In this approach, the LBEs are chosen by grouping similar accident sequences into an event class. Similar accident sequences are those that have similar initiating events and display similar accident behavior in terms of system failures and/or phenomena and lead to similar source terms. With LWRs for example, similar accident sequences would be events such as Anticipated Transient Without Scram (ATWS), various Loss of Coolant Accidents (LOCAs) (of different break sizes) with similar equipment response, containment bypass, transients of various types where each type exhibits similar equipment response. Similar accident sequences are also likely to have the same SSCs credited for accident prevention and/or mitigation. What are considered 'similar' groupings will be determined on a technology-specific basis.

The goal of the grouping process is to account for all the event classes with 95th percentile frequency greater than 1E-7 per year, and to strike a reasonable balance between the number of event classes and the degree of conservatism used in the grouping process. As a result of the grouping process, all PRA sequences are covered by an LBE. Sequences resulting in small doses can be covered with a few 'high' frequency LBEs, representing general event classes, that still satisfy the F-C curve. Higher dose sequences can be covered with more numerous LBEs, representing more detailed event classes, to show that they satisfy the F-C curve and associated criteria.

The main reason for including rare events is to ensure that no potentially high-consequence event is excluded due to the uncertainty in frequency alone. This approach provides additional confidence in the robustness of the design to withstand low-frequency, high-consequence events for risk goals (such as the QHOs).

Step 6 **Select an event sequence from the event class that represents the bounding consequence.**

The selected event sequence defines the accident behavior and consequences for the LBE that represent this event class. If several events within the event class have similar consequences, then a bounding event is selected. If there is not a clear bounding event, then the event with the lowest frequency is selected. The frequency of the event class is determined separately from the bounding consequence event. See Step 7.

Step 7 **Establish the LBE's frequency for a given event class.**

The frequency of an event class is determined by setting the LBE's mean frequency to the highest mean frequency of the event sequences in the event class and its 95th percentile frequency to the highest 95th percentile frequency of the event sequences in the event class. The mean and 95th percentile frequencies can come from different event sequences.

Step 8 **Verify that each LBE meets the probabilistic and deterministic acceptance criteria.**

The LBEs have to meet the F-C curve plus the defense-in-depth requirements that are a function of the LBE frequency range, as described below in Subsection 6.4.2 and summarized in Table 6-3. The measures needed to verify that the acceptance criteria are met are detailed in the example application in Appendix E. If criteria are not met, then either the event class is refined or modifications are made to the design.

Step 9   Repeat these steps for all event classes.

These steps should be repeated until all event classes have been considered.

As noted above, demonstrating compliance with the F-C curve at low dose levels via a PRA may be overly burdensome. An alternative LBE selection process may be needed for the high-frequency-low- dose region (i.e., events with frequencies ≥ 1E-3 per year). For this region, engineering judgment or experience (based, for example, on the event that is regarded as the most challenging for SSC design criteria and plant safety) and knowledge of the physics of the design may be used to select the LBEs for each class of events, for reasons of practicality.

Since the PRA being used is a maintain PRA, LBEs can change if the PRA changes during the life of the plant. Chapter 7 states requirements related to the living PRA.

For the LBE selection the question remains at what 'level' are the selected sequences defined: cut-set, systemic, or functional? In the Framework approach, the LBEs are sequences selected from the PRA at the 'systemic' level in terms of front-line systems that provide the needed safety functions. The specific level of detail for these 'front-line' systems for different technologies will be determined in the technology-specific Regulatory Guides.

Figure 6-4 shows how a particular set of event classes that starts out with similar failures may be broken down into 3 different classes and related LBEs and compared against the F-C curve.

The LBE associated with a relatively high frequency of 5E-2/yr has to have, by design, little consequence in terms of dose at the boundary. The LBE sequence with a lower frequency of 5E-4/yr can have higher consequences, as shown, and the LBE with 3 failures can have even higher consequences yet, but a lower frequency of 5E-6/yr. As stated above, other sequences besides the ones shown in Figure 6-4, which belong to the same event class, will contribute in terms of frequency to the LBE frequency of that class.

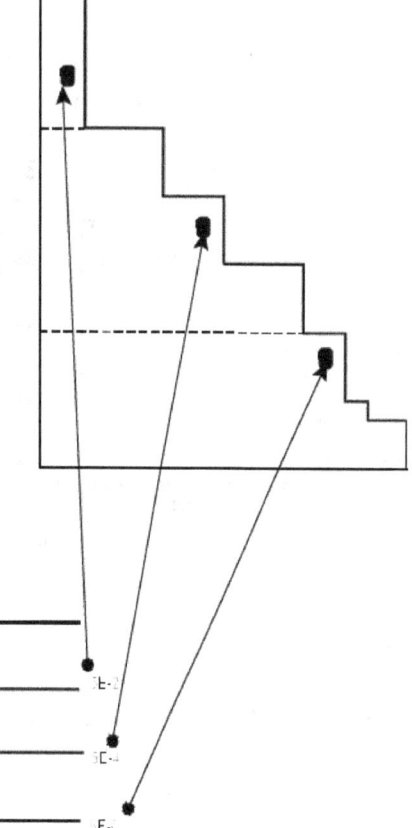

Figure 6-4   LBE Selection Schematic Example

Appendix E gives a detailed example application of the selection process.

6.4.2 Criteria to be Met by the LBEs

In addition to conforming to the F-C curve in the stringent manner described above, some additional criteria are imposed on the LBEs for defense-in-depth purposes and to ensure that the LBE criteria are risk-informed rather than risk-based. These criteria involve (1) additional deterministic criteria, and (2) additional dose criteria. Criteria are also imposed on initiating events. Finally, a deterministically chosen LBE is added to the LBEs obtained from the PRA. The LBE frequency ranges and all these criteria are discussed next.

6.4.2.1 Binning Probabilistic LBEs by Frequency

The additional criteria the LBEs have to meet, depend on the frequency range the LBEs fall into. The criteria should be most demanding at high frequencies , and become more relaxed at the lower frequencies. Consequently, the region under the F-C curve can be divided into frequency categories for purposes of specifying frequency-related deterministic criteria. Table 6-2 lists the proposed categories and their basis. The criteria for selecting the frequency categories take into account those events that are expected to

- Occur during the lifetime of a plant,
- Occur during the lifetime of the population of plants,
- Challenge the Commission's Safety Goals.

Table 6-2   LBE Frequency Categories

Category	Frequency (mean)	Basis
frequent	$\geq 10\text{-}2/ry$	Capture all event sequences expected to occur at least once in lifetime of a plant; assume lifetime of 60 years
infrequent	$< 10\text{-}2/ry$ to $\geq 10\text{-}5/ry$	Capture all event sequences expected to occur at least once in lifetime of population of plants; assume population of 1000 reactors
rare	$<10\text{-}5/ry$ to $\geq 10\text{-}7/ry$	Capture all event sequences not expected to occur in the lifetime of the plant population, but needed to assess Commission's safety goals

The frequencies in Table 6-2 apply to the entire event sequences, not the initiating event (IE) frequencies.

6.4.2.2 Additional Deterministic Criteria

For defense-in-depth purposes additional deterministic criteria are imposed on the LBEs. These additional criteria reflect some of the considerations that are found in the General Design Criteria of Appendix A to 10 CFR 50 for the current reactors. The additional deterministic criteria imposed on the LBEs vary with the frequency range.

In the frequent range, LBE frequency greater than or equal to 1E-2 per year:

• No impact on fuel integrity or reactor lifetime and the safety analysis assumptions occurs,
• No barrier failure occurs (beyond the initiating event),
• Redundant means of reactor shutdown remain functional,
• Redundant means of decay heat removal remain functional.

In the infrequent range, LBE frequency greater than or equal to 1E-5 per year, but less than 1E-2 per year:

• A coolable geometry is maintained,
• At least the containment functional capability remains,
• At least one means of reactor shutdown remains functional,
• At least one means of decay heat removal remains functional.

For the rare range, LBE frequency less than 1E-5 per year, no additional deterministic criteria apply.

6.4.2.3   Dose Criteria

The frequency ranges of the LBEs also affect some dose criteria the LBEs have to meet.

As noted in the F-C curve discussion of Section 6.3.1, the dose limits specified by Appendix I of Part 50 and by Part 20 are cumulative (over one year) dose limits. Therefore, in addition to meeting the per-event F-C curve, another requirement imposed for the LBEs with frequencies greater than 1E-3 per year is that they meet the cumulative dose requirements. This means a frequency weighted summing of the doses of all the LBEs in the range, as identified below. All these criteria are summarized as follows:

In the frequent range, LBE frequency greater than or equal to 1E-2 per year:

• The cumulative dose at the exclusion area boundary (EAB) meets the 5 mrem dose specification

In the infrequent range, LBE frequency greater than or equal to 1E-5 per year, but less than 1E-2 per year:

• The cumulative dose at the EAB of LBEs with frequencies greater than or equal to 1E-3 per year, meets the 100 mrem specification.

• For LBEs with frequencies less than 1E-3 per year, the worst (maximum based on meteorological conditions) two-hour dose at the EAB, and the dose from the entire duration of the accident at the outer boundary of the of the low population zone (LPZ), meets the F-C curve.

For the rare range, LBE frequency less than 1E-5 per year:

• The worst (maximum based on meteorological conditions) two-hour dose at the EAB, and the dose from the entire duration of the accident at the outer boundary of the of the low population zone (LPZ), meets the F-C curve.

To carry out the dose calculations, sequence-specific source terms are used for the consequence analysis associated with the probabilistic LBE sequences, based on the following criteria:

- They are selected from the design-specific PRA with due consideration for uncertainty,

- They are based on analytical tools that have been verified with sufficient experimental data to cover the range of conditions expected and to determine uncertainties,

- They reflect the sequence-specific timing, energy, form, and magnitude of radioactive material released from the fuel and coolant. Credit may be taken for natural and/or engineered attenuation mechanisms in estimating the release to the environment, provided there is adequate technical basis to support their use, and

- The radionuclide release fractions used to characterize the source term are the 95% value of the probability distribution.[15]

These criteria provide a flexible, performance-based approach for establishing sequence-specific licensing source terms. However, the burden is on the applicant to develop the technical bases, including experimental data, to support their proposed source terms. Specifically, the use of sequence-specific source terms requires the applicant to do sufficient testing to confirm the magnitude and rate of release, the timing and energy of release, the chemical form, and transport properties of fission products from the fuel, reactor coolant system, and reactor building under the range of conditions analyzed in the PRA. This includes accounting for the impact of different burn-up levels that the fuel can experience, and the physical and chemical conditions associated with various accident sequences on the release fractions and release rates of major fission product groups. Applicants can propose to use a conservative source term for LBEs, provided the use of such a source term does not result in design features or operational features that detract from safety.

6.4.2.4   Criteria on Initiating Events

In this document's approach there are also limits on the initiating event (IE) frequencies in the various frequency categories. To ensure that the assumptions in the PRA on initiating events (IEs) are preserved, each applicant proposes cumulative limits on IE frequency for each of the LBE event frequency categories. The cumulative initiating event limits are to be agreed upon between the applicant and the NRC consistent with the technology and safety characteristics of the design. These limits will be monitored during the plant operation as part of the maintain PRA program. Chapters 7 and 8 state the requirements related to the maintain PRA..

6.4.3   Deterministic Selected LBE

Chapter 4 describes a defense-in-depth approach that protects against unknown phenomena and threats, i.e., to compensate for completeness uncertainty affecting the magnitude of the source term from an accident. The approach ensures that regardless of the features incorporated in the plant to prevent an unacceptable release of radioactive material from the fuel and the reactor coolant system (RCS), there are additional means to prevent an unacceptable release to the public should such a release that has the potential to exceed the dose acceptance criteria occur.

---

[15]The upper value of the 95% Bayesian probability interval

Accordingly, as a deterministic defense-in-depth provision, each design needs to have the capability to establish a controlled low-leakage barrier if plant conditions result in the release of radioactive material from the fuel and reactor coolant system in excess of anticipated conditions. The specific conditions for the barrier leak tightness, temperature, pressure, and time available to establish the low-leakage condition will be design-specific. The design of the controlled leakage barrier should be based upon a process that defines an event representing a serious challenge to fission product retention in the fuel and coolant system. This event should be agreed upon between the applicant and the NRC consistent with the technology and safety characteristics of the design. The event could represent an assumed fuel damage event, such as a graphite fire in a High Temperature Gas-cooled Reactor (HTGR). The controlled leakage barrier provides a fission product containment and, as such, is a policy issue needing the Commission's review and direction. Appendix C discusses this issue further.

The deterministic LBE is to be analyzed mechanistically to determine the timing, magnitude, and form of radionuclide released into the reactor building, and the resulting temperature, pressure and other environmental factors (e.g., combustible gas) in the building during the event. The timing of closure and the allowable leak rate is then established such that the worst two-hour dose at the EAB and the dose at the outer edge of the low population zone (LPZ) for the duration of the event do not exceed 25 rem TEDE. The deterministic LBE is also used for siting purposes.

A dose of 25 rem is the current off-site dose guideline for design basis accidents in 10 CFR 50.34 and 10 CFR 100. It is also the dose that defines an abnormal occurrence (AO), as described in the Commission's April 17, 1997, Policy Statement on AOs [NRC 1997], which specifies substantial radiation levels to imply a whole-body dose of 25 rem to one or more persons.

6.4.4 Summary of the Risk-Informed Licensing Process

This report discussed several risk-informed features that are included in the licensing approach, as envisioned in the Framework. The most important features are summarized below.

Acceptable Plant Risk

- In the Framework, the criterion for an acceptable integrated plant risk is compliance with the NRC safety goals and their associated Quantitative Health Objectives (QHOs).

- Further requirements could be specified in the future for the total risk profile, for example, in terms of an acceptable complementary cumulative distribution function (CCDF) for a risk measure of interest, e.g., early fatalities. Such a CCDF would provide a regulatory limit for the appropriate risk measure as a function of frequency.
- A single limiting criterion on consequence or its surrogate (such as CDF or LERF for LWRs) cannot be developed from diverse reactor designs..

- While not representative of integrated plant risk, a set of probabilistically selected bounding Licensing Basis Events (LBEs) provide added confidence that the plant risk is acceptable.

Frequency-Consequence Curve

- The purpose of the Frequency Consequence (F-C) curve in this NUREG is as a tool for selecting LBEs in a risk-informed manner and to provide a criteria for LBE acceptability.

- To meet the purpose of LBE selection, the F-C curve in this NUREG is a 'per event' curve, not a cumulative measure of total risk.

- The proposed curve has some anchor points based on existing dose requirements in the regulations, but the existing regulations do not relate these doses to frequencies. Engineering judgement is used to relate specific dose ranges to specific frequency ranges in the proposed curve.

- Other variations of the F-C curve could be proposed with similar arguments. A single 'per event' curve could be replaced by two curves: (1) one in the high-frequency-low-consequence range, based on cumulative (annual) requirements in the current regulations, which plots rem/yr vs probability, and (2) a curve in the lower frequency-higher consequence range, based on dose limits for individual accidents in the current regulations.

Licensing Basis Event Selection and Criteria

- The LBEs are selected in a risk-informed manner from the PRA sequences, and the LBEs need to meet the proposed F-C curve.

- Each LBE represents a class of PRA sequences, and each LBE bounds the sequences in its class in terms of frequency and consequence.

- Collectively, LBEs bound all the individual PRA sequences in frequency and consequences.

- In the high-frequency-low dose range, LBEs may have to be chosen based on judgement rather than PRA results.

- In addition to conforming to the F-C curve, some other criteria are imposed on the LBEs. These criteria involve (1) additional deterministic criteria, and (2) additional dose criteria. The severity of these additional requirements depends on the frequency range the LBEs fall into. Some criteria on initiating events are also imposed.

- The LBEs need to meet the F-C curve and the additional criteria using only the credited safety-significant SSCs

- As a defense-in-depth provision, a deterministic LBE is added that ensures each design can establish a controlled low-leakage barrier if plant conditions result in the release of radioactive material from the fuel and reactor coolant system in excess of anticipated conditions. This event should be agreed upon between the applicant and the NRC consistent with the technology and safety characteristics of the design.

- Table 6-3 summarizes the criteria imposed on the probabilistically selected LBEs.

Table 6-3   LBE Criteria

Frequency Category	Additional acceptance criteria for LBEs (demonstrated with calculations at the 95% probability value* with success criteria that meet adequate regulatory margin, as discussed in Section 6.6)
frequent ($\geq 10^{-2}$)	• no barrier failure (beyond the initiating event) • no impact on fuel integrity or lifetime and safety analysis assumptions • redundant means for reactor shutdown and decay heat removal remain functional • annual dose to a receptor at the EAB $\leq$ 5mrem TEDE
infrequent ($< 10^{-2}$ to $\geq 10^{-5}$)	• maintain containment functional capability • a coolable geometry is maintained • at least one means of reactor shutdown and decay heat removal remains functional • for LBEs with frequency > 1E-3 annual dose to a receptor at the EAB $\leq$ 100mrem TEDE • for LBEs with frequency < 1E-3 the worst two-hour dose at the EAB, and the dose from the duration of the accident at the outer boundary of the LPZ, meet the F-C curve
rare ($<10^{-5}$ to $\geq 10^{-7}$)	• the worst two-hour dose at the EAB, and the dose from the duration of the accident at the outer boundary of the LPZ, meet the F-C curve
Note: With the exception of the source term, realistic calculations are carried out to obtain the mean and uncertainty distribution of the important parameters for estimating frequency and consequences. Source Term calculations use the 95% probability value* of the amount of radionuclides released, obtained from a mechanistic calculation, based on validated analytical tools as discussed in Section 6.4.2.3. Dose calculations use RG 1.145 or the equivalent for calculating atmospheric dispersion.[16]	
EAB - exclusion area boundary LPZ - low population zone (as defined in 10 CFR 100) TEDE - total effective dose equivalent * The upper value of the 95% Bayesian probability interval	

Safety-significant SSCs

•   The SSCs credited in LBEs to achieve compliance with the LBE criteria are risk-significant and require special treatment.

•   A plant designer has flexibility in choosing a set of SSCs to credit in the LBE calculations, i.e., which set is safety-significant.

•   The type of special treatment an SSC receives depends on the function and importance of the SSC. The treatment required should ensure that the SSCs will perform reliably, as postulated in the PRA, under the conditions (temperature, pressure, radiation assumed to prevail in the event sequences for which the SSC's successful function is credited in the risk analysis.

---

[16]Further guidance will be provided on how to calculate source terms at the 95% probability value.

6.5   Safety-significant SSCs and Special Treatment

The aim of this document is to incorporate a safety classification scheme in which all the plant SSCs fall into two categories, safety significant or non-safety significant, distinguished by whether the SSCs need special treatment or not.[17] As discussed under the LBE selection process, the term 'safety-significant' is assigned to those SSCs whose functionality plays a role in meeting the acceptance criteria imposed on the LBEs. These SSCs require special treatment.

The term 'special treatment' is used to designate requirements imposed on SSCs that go beyond industry-established requirements for equipment classified as "commercial grade." These requirements provide additional confidence that the equipment is capable of meeting its functional requirements under PRA-analyzed conditions. The type of special treatment varies dependent on the function and importance of the SSC. The treatment helps to ensure that the SSCs will perform reliably (as postulated in the PRA) under the conditions (temperature, pressure, radiation,) assumed to prevail in the event sequences for which the SSC's successful function is credited in the risk analysis.

A basic special treatment requirement for all safety-significant SSCs will be the establishment and monitoring of reliability and availability goals. All safety-significant SSCs will have reliability and availability consistent with the values assumed in the PRA. During operation, a process similar to the monitoring of the performance and condition of structures, systems, or components, against licensee-established goals of 10 CFR 50.65, the Maintenance Rule, is expected to be an integral part of the monitoring program for this special treatment requirement. Monitoring will consist of periodically gathering, trending, and evaluating information pertinent to the performance, and/or availability of PRA-related SSCs, and comparing the result with the established goals and performance criteria to verify that the goals are being met. When the goals are met, the plant's performance is consistent with the licensing bases. When a goal or performance criteria are not met, then assessment of the affect of the performance issue on the PRA and licensing bases is required. Cause determination and corrective actions may also be required. Performance issues that result in the failure of Framework acceptance criteria will require licensing action.

Other special treatment requirements may be required depending on the function and importance of the SSC. Risk importance calculations could be used to focus the types of special treatments that would be applied to an SSC. However, these importance calculations would have to use other risk measures than the CDF and LERF used for LWR calculations. For calculating SSC risk importance based on the F-C curve, a process like the following could be used:

- From the PRA that is carried out as part of the design of the plant, all sequences that can potentially result in a dose at the site boundary greater than a certain selected dose, for example $\geq$ 1 rem, are selected.

- Next, an importance measure or measures (IM) are defined. These measures can be defined analogous to importance measures related to core damage frequency (CDF), or large early release frequency (LERF) for LWRs, i.e., analogous to risk achievement worth, RAW, risk reduction worth, RRW or the Fussell-Vesely, F-V. This IM can be based, for example, on the notion of a "1 rem exceedance frequency" (EXF[1]), i.e., the sum of frequencies of all sequences that exceed the criterion of a 1 rem dose at the site boundary.

---

[17]This differs from the scheme for current reactors, where distinction is made between 'safety related,' 'safety significant,' and 'important to safety' equipment.

- Then the value(s) of the IM related to EXF[1] is calculated for all SSCs that appear in the sequences that result in doses greater than 1 rem. These are trial values of IM. If desired, additional weight can be given in the definition of IM to SSCs that appear in (lower frequency) sequences leading to higher values of dose at the site boundary, e.g., in excess of 25 rem, 50 rem, or 100 rem as displayed on the F-C curve. This would involve calculating analogs of IM that are related to EXF[25], EXF[50], EXF[100], etc., and then choosing appropriate weights for the values of IM that result, to calculate the total value of the importance measure. Such a process could be used also to emphasize risk aversion, i.e., assigning importance to SSCs in proportion to the dose that the SSC prevents or mitigates.

Such a process may be useful for providing graded treatment to the safety-significant SSCs related to their importance measures. The type of treatment that safety-significant SSCs receive is determined from the conditions that the SSC is assumed to operate under, based on the sequences where it is needed. Verification of the functionality of the SSC under the required conditions is demonstrated via a reliability assurance program.

However, that in calculating the average individual risk of early fatality and latent cancer fatality for comparison with the QHOs of the Safety Goal Policy, all the systems modeled in the PRA can be credited in the risk calculations, not just the safety-significant systems credited in the LBE calculations.

6.6   Safety Margin

Safety margin is incorporated in the framework as an element of defense-in-depth through the use of requirements containing regulatory margin and by encouraging the designer to include an 'operational' margin as an element of the design. Regulatory and operational margin were introduced in Section 4.6. The purpose of this section is to describe how these two types of safety margins are addressed within the framework.

In keeping with the treatment of a draft NUREG on the subject of safety margins[18], it is assumed that the operation of a safety system or barrier can be characterized by one or more safety variables, e.g., pressure, temperature, ductility, and that the safety variables can demarcate the transition from "intact" to "lost function." If the safety variable remains in an acceptable range, then the system or barrier remains functional. If the variable exceeds that range, then loss of function occurs. These safety variables need to be directly or indirectly measurable, and their values need to be predictable for plant conditions during normal and emergency operation.

As shown in Figure 4-4, for purposes of the Framework, the safety margin is the sum of (1) the regulatory margin, which is the difference between the ultimate capacity of the safety variable and the regulatory limit of the safety variable, and (2) the operational margin, added at the option of the designer, which is the difference between the regulatory limit of the safety variable and the value of the safety variable at which the system or barrier is expected to perform, according to the design. The draft NUREG on safety margins states, "Adequate safety margins" are inextricably linked to safety limits—limiting values imposed on safety variables (e.g., peak clad temperature (PCT) and containment pressure in current LWRs). Thus, when operating conditions stay within safety limits, the barrier or system has a negligible probability of loss of function, and an adequate safety margin exists.

---

[18]NUREG-1870 is anticipated to be published in 2008.

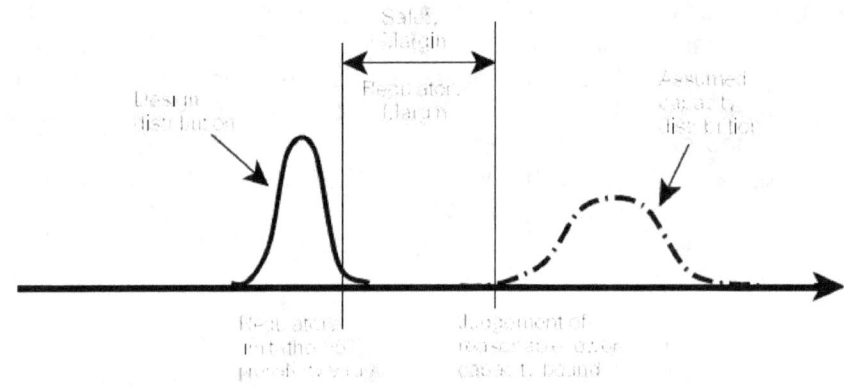

Figure 6-5   Safety Margin

In practice, the design value and the ultimate capacity of a safety variable are not particular single values but probabilistically distributed quantities. Therefore, one has to specify (1) from where on the capacity distribution the regulatory limit should be measured from, and (2) what part of the design analysis distribution should be used to show compliance with the regulatory limit, and (if pertinent) show additional margin beyond that provided by the regulatory limit.[19] This question is further complicated by the fact that accurate distributions, especially for the ultimate capacity of the safety variable, are often not available. Such information may be beyond the current state-of-the-art or very costly.

This lack of detailed information about the capacity distribution is often addressed by picking a safety variable capacity value that reflects judgement as to where the onset of damage of the system or barrier being analyzed occurs. This value reflects the judgement of what is a reasonable bound on the lower limit of the capacity distribution, given its assumed uncertainty. Further judgement then is used to set the safety limit below this onset of damage value by an amount that is commensurate with the lack of data, and the importance of the safety variable, and the system or barrier whose functionality it determines. This judgement reflects an allowance for unknowns that are part of the incompleteness uncertainty.

The design analysis distribution of the safety variable is usually less costly to obtain and better known. This NUREG advocates using the 95% probability value (The upper value of the 95% Bayesian probability interval) of the design distribution to show that the regulatory limit is met. Figure 6-5 shows these concepts.

In Figure 6-5, the capacity distribution, which is not well defined, is indicated by a dashed line. Judgement is used to pick a capacity value, below which the probability of failure of the system or barrier whose operation is characterized by the safety variable is considered negligible. This represents a lower bound on the needed capacity. If the safety variable exceeds this value, loss of function is assumed. The regulatory margin is then the distance between this lower bound capacity value and a point on the upper tail of the design distribution of the safety variable. A point corresponding to the 95% probability value on the tail of the design distribution is suggested as a

---

[19] If the probability distributions of both design analysis and capacity were perfectly known and included both random and state-of-knowledge uncertainties, the distributions could be convoluted and the probability of failure (i.e., design value exceeds capacity) obtained directly from the convolution. If this probability was not low enough, the design could be adjusted until it was, and there would be no need for safety margin. However, in reality the distributions are not perfectly known.

credible point on the distribution, if the distribution is reasonably well established. As noted above, the judgement in setting the regulatory margin will be influenced by (1) the availability of data about the safety variable distributions, (2) the importance of the safety variable in characterizing the safe operation of the system or barrier, and (3) the importance of the system or barrier itself. The intent is to allow margin for phenomena and processes that may have been inadequately considered in generating models to simulate the behavior of the given system or barrier. The margin thus provides an allowance for unknowns that are part of the incompleteness uncertainty.

In Figure 6-5 the safety margin and regulatory margin are identical, i.e., no additional operational margin is deliberately provided. Figure 6-6 shows the situation where the designer has added the additional margin to further increase the safety margin.

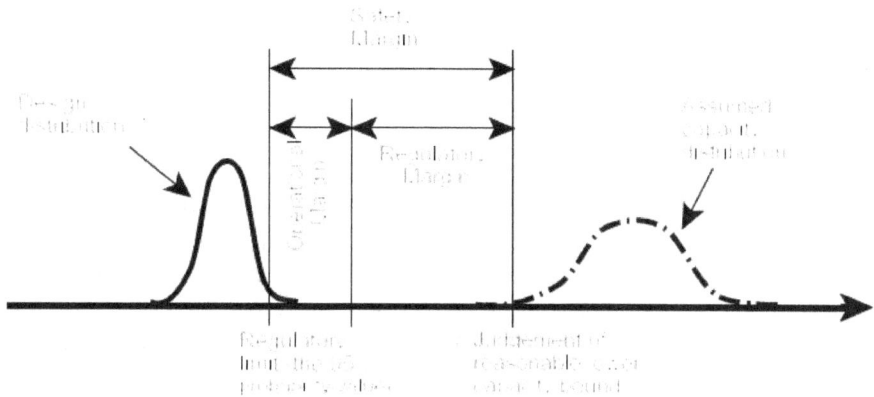

Figure 6-6   Safety Margin with Operational Margin

Chapter 4, the operational range of the safety variable, represented here by the design distribution, has been moved to lower values to provide an operational margin to add to the safety margin. This may be included by the designer to ensure regulatory limits are easily met even if, e.g., the operational range of the safety variable I changes in the future.

Based on either Figure 6-5 or 6-6, the safety margin for the safety variable can be taken to be the distance between the bounding prediction of the design analysis probability distribution of the safety variable and the point at which failure becomes non-negligible on the capacity probability distribution of the safety variable. The phrases "sufficient margin" and "adequate margin" can be taken to mean that regulatory margin is preserved.

6.6.1   Regulatory Safety Margin in the Framework

Safety margin is incorporated into the Framework as an element of defense-in-depth. Since the Framework is performance-based, it does not prescribe specific margins at the various levels of a safety analysis.

6.6.1.1   Frequency-Consequence Curve

At the highest level of the safety analysis, involving the results of the PRA and the LBE calculations, the Framework establishes the regulatory limits of the frequency-consequence curve consistent with the discussion above. The F-C curve limits are defined at the different frequencies by dose levels which reflect the regulatory limits on dose at the frequencies specified. These limits are based on conservative decision-making rooted in current practices and reflect judgement as to the acceptable risk for a given likelihood and consequence. The F-C curve is an example of a regulatory limit where the capacity distribution is not well understood and where judgement was used to set the safety limit below the onset of unacceptable risk.

The margin associated with the F-C curve results in both a consequence margin and a frequency margin. Consequence margin is the difference between the predicted dose and acceptance dose for a given LBE frequency. It is determined by comparing the predicted dose to the applicable acceptance dose on the frequency-consequence curve. Frequency margin is the difference between the calculated frequency and the frequency used to establish the acceptance criteria and the defense-in-depth requirements.

As indicated in the earlier discussions, these LBEs have to meet the F-C curve at the 95% probability level in both frequency and consequence. The use of the 95th percentile is consistent with the margin discussion above.

For consequence margin, the probabilistic requirements for the LBEs dose consequence calculation are determined using the 95th percentile value of the source term. For frequency margin, an LBE's frequency is set to the highest 95th percentile frequency of the event sequences in the event class. This approach of using the 95th percentile values and regulatory margin associated with the F-C curve increases the evaluated frequency of the LBE and tightens the acceptance criteria, and potentially results in a possible shifting of the LBE into a higher frequency category associated with a lower maximum acceptance dose and more restrictive deterministic requirements.

In addition, the PRA event sequence frequencies are influenced by their success criteria and by the selection of reliability and availability goals, for which margins are discussed below. For reliability and availability goals, the designer may choose to include operational margin within these goals, which will be reflected in the overall frequency of each event sequence.

6.6.1.2 Safety Variable Limits

In addition to the margin included in the F-C curve, regulatory margin could be specified for selected key variables on a design-specific basis. This includes key variables used in determining PRA success criteria.

Within the PRA, success criteria are used to distinguish the path between success and failure for components, human actions, trains, systems, structures and sequences. In all cases, the success criteria are to be fully defensible and biased such that issues of manufacturer or construction variability, code limitations and other uncertainties are unlikely to result in a failure path being considered a success path. Ensuring that success paths are truly success paths could be supported by requiring regulatory margin for selected key variables and by encouraging the incorporation of operational margin.

6.6.1.3 Code and Standards

This document does assume that appropriate codes and standards are used for the analyses of systems, structures, and components for determining the plant's level of safety, and that the margins in these analyses follow the general guidelines presented here. Specific guidance is not provided for codes and standards within the Framework for it is expected that most codes and standards will be associated with a design-specific features.

6.6.1.4 Completeness

This NUREG includes a process for identifying a complete set of probabilistic LBEs that bounds all PRA event sequences having a 95th percentile frequency greater than 1E-7 per year.

Identifying a complete set of LBEs is key to ensuring that adequate safety margin is achieved as it provides the starting point for which safety variables and codes and standards are applied.

The scope of the PRA used for this document encompasses the whole spectrum of events that credibly occur during the life of the plant: normal operation, as well as frequent, infrequent, and rare initiating events and accident event sequences. This scope is broader than that used currently for LWR risk analysis, which concentrates on beyond design basis accidents, i.e., accidents leading to severe core damage, and uses LWR-specific surrogate metrics like core damage frequency (CDF) and large early release frequency (LERF).

This document also includes a deterministic LBE that addresses the limiting challenge to the final radiological barrier to address uncertainties that are not fully known.

The use of a design-specific PRA to identify a bounding set of LBEs ensures that unique accident behavior and phenomena are assessed. The use of a deterministic LBE for assessing the challenge to the final radiological barrier adds margin for completeness or knowledge uncertainty. The use of a deterministic LBE is one of the defense-in-depth measures that the Framework requires to address incompleteness uncertainty.

6.6.2 Operational Margin

The designer can incorporate an additional margin, called the "operational margin" in the Framework, by designing a system so it operates below the regulatory limit for normal operations and excursions. This NUREG encourages the use of margin beyond that required, to minimize changes in the licensing bases as the result of identifying new licensing bases events that may result from unexpected changes in performance or knowledge. This operational margin can be applied to the LBE consequence analysis to create margin between these results and the limits of the F-C curve, to safety variables used in success criteria and codes and standards, and to the reliability and availability goals established for the SSCs within the PRA.

For example, the operational margin incorporated in the reliability and availability goals could vary depending on the SSC's function, performance uncertainty, or importance. These goals should be designed such that variations in SSC performance are unlikely to change the selection and characteristics of the LBEs.

The incorporation of operational margin into the various elements of the Framework should help to provide added insurance that plant operation and event response will not deviate from the inputs and assumptions and the associated analysis used to demonstrate compliance with the Commission's safety goals.

6.7   Security Performance Standards

The purpose of this section is to define proposed risk-informed and technology-neutral security expectations and performance standards for new proposed plants.

Requirements related to security for current Nuclear Power Plants (NPPs) are contained in 10 CFR 73 and in post 9/11 orders. These requirements are primarily prescriptive and require plant-specific assessments, features, and measures related to protection against the design basis threat (DBT). These requirements cover several subjects, including

- guard force and training
- protection of vital areas
- personnel, package and vehicle screening
- security fences and detection devices
- mitigation strategies
- cyber protection

These requirements provide a baseline of preventive and mitigative features directed toward protecting public health and safety, the environment, and the common defense and security from the DBTs related to sabotage, armed intrusion, external attack, and theft or diversion of nuclear material. They generally require measures to detect, delay, assess, and respond to security-related threats up to and including the DBT. For advanced reactors, the staff proposed (in SECY-07-0167 [NRC 2007]) a revision to the Commission's 1986 Advanced Reactor Policy statement to include security considerations. This proposed revision states that for advanced reactors, security expectations are that they will provide enhanced margins of safety and use simplified, inherent, passive, or other innovative means to accomplish their safety and security functions.

6.7.1   Security Expectations

In SECY-05-0120 [NRC 2005a], the staff proposed that security performance standards for Generation IV and other future reactor concepts be developed as part of the Framework. In a September 9, 2005, SRM [NRC 2005b], the Commission approved the staff recommendation and requested that expectations for security be integrated with safety and preparedness and that security-related design issues be resolved at the early stage of the regulatory review process, so that there will be less reliance on operational programs. Accordingly, a proposed set of expectations that defines risk-informed and performance-based security for new plants was developed as discussed below. These security expectations describe, in qualitative terms, what security at nuclear power plants should achieve. They address the level of safety and security that should be achieved, the scope of what should be protected and considered, and key aspects of the approach that should be followed. They also provide guidance on the scope and purpose of proposed security performance standards, which are discussed in Section 6.7.2.

Specifically, the proposed security expectations for new plants encompass the following:

- Protection of public health and safety, the environment, and the common defense and security with high assurance is the goal of security.

- The overall level of safety to be provided for security-related events should be consistent with the Commission's expectations for safety from non-security related events.

- Security should be considered integral with (i.e., in conjunction with) safety and preparedness.

- A defined set of events outside the DBT should be considered, as well as the DBT, to identify vulnerabilities, and provide margin.

- Defense-in-depth should be provided against the DBT and each event outside the DBT considered, to help compensate for uncertainties.

- Security should be accomplished by design; as much as practical.

These expectations define the elements which should be addressed in the security performance standards. These security expectations are intended to promote enhanced security, emphasize design solutions to security issues, provide means to ensure integration of security, safety and preparedness, and provide guidance for qualitative and quantitative measures for assessing security. They are intended to be consistent with existing measures of safety (e.g., public health and safety) wherever possible, and relevant aspects of NRC's safety philosophy (e.g., defense-in-depth). In addition, the security expectations and performance standards are intended to apply to on-site fresh and spent fuel storage, as well as the reactor. Also, it was assumed in developing these expectations, that the design would be in compliance with 10 CFR 73 requirements and the post 9/11 orders. The security performance standards and their basis are described in the next section.

6.7.2 Security Performance Standards

The goal of the security performance standards is to define quantitative and qualitative criteria that can be used to determine whether the security expectations discussed above are met. The focus of the security standards is to be on performance of the system, rather than on prescriptive requirements. As such, the performance standards developed are consistent with the guidance in NUREG/BR-0303 "Guidance for Performance-Based Regulation." The guidance in this NUREG identifies the characteristics which should be present to have performance-based requirements, which includes

- Can observable characteristics, together with objective criteria, provide measures of safety performance?

- What is the performance level desired?

- Can corrective action be taken if the level of performance is not met? and

- Is flexibility for NRC, applicants and licensees provided?

Accordingly, the security performance standards should conform with this guidance.

To assess performance, each new plant should be required to conduct a safety and security assessment and evaluate the results against the performance standards described in this section. This safety and security assessment is intended to:

- Consider safety and security integral with design, rather than as post-design compensatory measures,

- Consider safety and security in developing preparedness measures, and

- Ensure that the relationship and affect among security, safety, and preparedness is considered in decision making.

The assessment should involve characterization of the potential threat, the potential targets, and the potential consequences. The adequacy of the methodology used would also need to be established. However, the likelihood of the threat depends on some factors only known to the adversary, so it is expected to be outside the scope of the assessments. Nevertheless, the types, objectives, and capabilities, as well as the strategies of the potential adversaries should be taken into account in the assessment.[20] The scope and guidelines for performing the safety and security assessment would be provided in a separate document.

Security needs should be considered during the design stage. Design, security, and preparedness decisions should be made in an integral fashion. During design, reasonable assumptions about the threats that need to be considered over the full life of the facility can be made, together with a spectrum of threats the facility is likely to encounter. It is important that these threat assumptions include the broad categories of potential physical protection challenges. These broad categories include

- Insider attack,
- Armed intrusion,
- Stand off attacks,
- Cyber attacks,
- Theft or diversion of nuclear materials,
- A credible combination of the above.

The theft or diversion of radioactive materials, namely fuel, is included in the broad category for potential physical protection challenges only for reactor designs that use mixed oxide (MOX)[21] or highly enriched uranium (HEU) fuel. Theft or diversion is not considered for reactor designs that use low enriched uranium (LEU) fuel. For fresh LEU fuel that has not been irradiated, subsequent dispersal would not create a significant hazard for public health and safety, the environment, and common defense and security, and it is not suitable for producing a weapon. For spent LEU fuel, its high level of radioactivity generally makes it self-protecting and difficult to steal. Therefore, theft or diversion of LEU nuclear fuel is not anticipated to be a significant contributor to risk at commercial nuclear power plants.

The threats that should be considered in the evaluation are the DBT and a defined set of events outside the DBT selected and categorized consistent with NRC safety and security assessment guidance for threat level severity (see Table 6-4). The events outside the DBT selected for evaluation should be sufficient for showing that the design has margin to withstand security-related events outside the DBT and for identifying any significant vulnerabilities or consequence thresholds.

---

[20]Detailed information about these threat descriptions is sensitive because access to this information would allow potential adversaries to better predict the capabilities of physical protection systems.

[21]MOX fuel utilizes weapons grade plutonium (4 -7 weight percent) in a depleted uranium matrix.

This will then help to compensate for uncertainties and also help provide high assurance of protection of public health and safety, the environment, and the common defense and security.

Table 6-4   Threat Level Severity

Level	Threat Description
High	Capability exists, intentions stated and history make this a likely credible threat.
Medium	Capability and history exist, but no stated intentions make this a likely threat.
Low	Capability exists, but no stated intentions or history make this an unlikely threat.
Negligible	Neither capability, intentions nor history exists and the threat is not considered credible or likely.

A risk-informed and performance-based approach was taken in developing security performance standards. This approach uses a combination of risk criteria to define the level of safety desired and deterministic criteria to complement the risk criteria to help account for uncertainties. An integrated decision process is then used to assess the various elements of the standards and the need for any additional action.

Risk information is useful in helping to make decisions on the level of protection to be provided against various events, and the relative importance of plant equipment, actions, or modifications in responding to those events. As such, having some probabilistic elements as part of the security performance standards would provide a means to judge the importance of various threats and potential solutions. However, since the probability of the threat itself has considerable uncertainty, any probabilistic performance standards would be most practical and useful if they are expressed as conditional (conditional upon the initiating event) probabilistic values and used in an integrated decision-making process that considers all factors.

The security performance standards proposed for new plants are as follows:

- Probabilistic Performance Standard:

    — Assess and take action on potential vulnerabilities to high, medium, and low level threats in accordance with Figure 6-7.

- Deterministic Performance Standards:

    — Ensure that the plant design, operation and security provide multiple layers of defense against each security-related threat that could endanger public health and safety, the environment, or the common defense and security.

    — Ensure that the plant design, operation, and security provide both prevention and mitigation measures for each security-related threat that could endanger public health and safety, the environment, or the common defense and security.

6. Licensing Basis

- Theft or Diversion Performance Standards:

   — For plant designs using MOX or HEU fuel, ensure that detection, delay, response measures and a surveillance program are provided consistent with 10 CFR 73 requirements.

- Design Solution Performance Standard:

   — The resolution of security-related issues should be effectively resolved through (1) facility design and engineered security features, and (2) formulating mitigation measures, with reduced reliance on human actions.

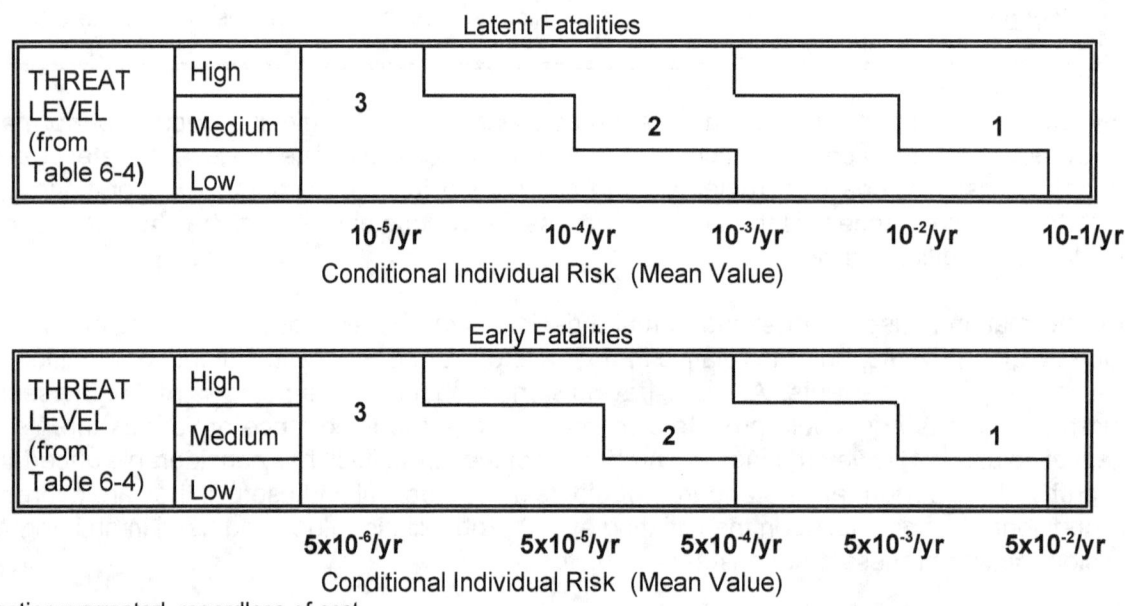

Latent Fatalities

THREAT LEVEL (from Table 6-4)	High			
	Medium	3	2	1
	Low			

10⁻⁵/yr   10⁻⁴/yr   10⁻³/yr   10⁻²/yr   10-1/yr
Conditional Individual Risk (Mean Value)

Early Fatalities

THREAT LEVEL (from Table 6-4)	High			
	Medium	3	2	1
	Low			

5x10⁻⁶/yr   5x10⁻⁵/yr   5x10⁻⁴/yr   5x10⁻³/yr   5x10⁻²/yr
Conditional Individual Risk (Mean Value)

1 = action warranted, regardless of cost
2 = cost-benefit region
3 = no action warranted

Figure 6-7   Conditional Risk

The technical basis for each of these standards is discussed below.

Probabilistic Performance Standard

The purpose of the probabilistic performance standard is to define the level of safety desired and, based upon this definition, identify potential vulnerabilities and the effectiveness of potential solutions to reduce those vulnerabilities. Since safety corresponds to the protection of public health and safety, the risk metrics used need to also be related to public health and safety as well as be developed from diverse reactor designs.. Accordingly, for consistency with previous Commission expectations for safety and with the overall plant risk described in Section 6.3, the quantitative health objectives (QHOs) from the Commission's 1986 Safety Goal Policy Statement were selected as the risk metrics to be used. These QHOs are expressed as individual risk of a latent fatality (2 x 10⁻⁶/yr) and an early fatality (5 x 10⁻⁷/yr) and are applicable out to 10 miles and 1 mile, respectively, from the exclusion area boundary of the plant. The use of the QHOs also provides a common metric to integrate safety, security, and preparedness.

To use the QHOs as an element of the security performance standards, the conditional risk (latent fatality and early fatality individual risk, assuming a probability of one for the initiating event) from each postulated security threat would need to be calculated and compared to a qualitative assessment of the severity of the threat level (see Table 6-4). Those threats judged high, medium, or low should be evaluated as potential vulnerabilities in accordance with Figure 6-7. The basis for the conditional risk values shown in Figure 6-7 comes from trying to maintain a level of safety equivalent to the QHOs. Since the probability of the threats occurring is not known, estimates of the range of probability of occurrence for each threat level were assigned as follows to support establishing the conditional risk values for early and latent fatalities shown in Figure 6-7, consistent with the QHOs:

> High — 0.01 to 0.001
>
> Medium — 0.001 to 0.0001
>
> Low — 0.0001 to 0.00001

As an example, the starting point of 0.01 was chosen considering that there are approximately 100 plants operating in the country and, therefore, the likelihood of any one or several simultaneously being a target is in the range of 1/100. If, in the future, more than 100 plants are operating, the 0.01 probability will represent a conservative value. In addition, for the purposes of this report, the QHOs themselves were reduced by a factor of 10 to account for the fact that more than one threat is being considered and, therefore, the cumulative risk from all threats is the measure of interest. Modifications which could reduce the conditional risk of the potential vulnerabilities should then be evaluated as shown in Figure 6-7. The regions shown in Figure 6-7 are intended to help ensure that the safety goal level of safety is achieved, when the cumulative effect of security-level threats is considered. Specifically, region 1 represents an area where the conditional risk is high enough to warrant action; region 2 represents an area where additional justification for action is needed and, thus, cost-benefit should be considered; and region 3 represents an area where the QHOs will likely be met and, therefore, no action is warranted. Accordingly, ensuring that vulnerabilities to threats that could result in high conditional risk are assessed in accordance with Figure 6-7 represents a probabilistic element of the security performance standards.

In a technology-specific application of this performance standard, a technology-specific set of risk metrics may be substituted for the QHOs (e.g., CDF and LERF for LWRs).

Deterministic Performance Standards

Any analysis of security-related events carries with it considerable uncertainty, mainly due to the uncertainty in likelihood of occurrence and consequences associated with the potential threats. Accordingly, to help compensate for this uncertainty, a defense-in-depth deterministic approach is proposed, such that the design should have multiple layers of defense against a range of postulated threats (i.e., defense-in-depth). These layers of defense can consist of preventive or mitigative measures, but should be consistent with the following principles:

- Plant security should not depend upon a single element of design, operation, maintenance, or security, and

- Plant design, operation, and security measures should address both prevention of threats and mitigation of potential consequences.[22]

These principles intend to ensure that the plant design, operation, and security provides multiple ways to defend against security-related threats. This can be accomplished by (1) having a plant that is not susceptible to a single action by an individual or group of individuals that threatens public health and safety, and (2) addressing both threat prevention and potential consequence mitigation. In addition, these principles integrate security with the basic concept of defense-in-depth used in reactor safety and preparedness, namely designing in measures to protect against accidents and ensuring mitigation is available if the accident happens. Each of these principles is discussed below along with its related performance standard.

For the first principle, the plant design, operation and security should include multiple ways of providing protection (e.g., multiple systems, barriers, response plans) in a security-related event. This is consistent with the fundamental approach and principles of defense-in-depth, which helps ensure that the plant safety is not dependent upon a single security measure for a threat that could adversely affect public health and safety, the environment, or the common defense and security. Accordingly, providing multiple layers of defense against each of the postulated threats represents an element of the security performance standards.

For the second principle, the design should include measures to both prevent and mitigate security-related events from being a threat to public health and safety. As such, the design, operation, preparedness, and security measures need to include at least one preventive and one mitigative response for each security-related threat. In this regard, prevention means preventing the security-related threats from causing the intended plant damage and mitigation means, assuming the intended damage occurs, what can be done to reduce its consequences and protect public health and safety. Accordingly, ensuring preventive and mitigative measures are provided for each threat represents an element of the security performance standards.

Theft or Diversion Performance Standard

For theft or diversion of MOX or HEU fuel, the risk to public heath and safety could be substantial. However, this risk would not likely be to the population surrounding the plant, but could be a hazard to an area far from the plant. Accordingly, the use of the Commission's Safety Goals to define the level of safety desired may not be appropriate. Therefore, ensuring new plant designs include sufficient detection, delay and response capability, and a surveillance program capable of rapidly detecting the loss of material (before a goal quantity can be lost) is an element of the security performance standards. In this regard, the requirements described in 10 CFR 73 should be sufficient.

Design Solutions Performance Standard

Consistent with Commission's Advanced Reactor Policy Statement, security-related issues should be effectively resolved through (1) facility design and engineered security features, and (2) formulating mitigation measures with reduced reliance on human actions. Accordingly, when alternative solutions to the resolution of security issues are possible, those alternatives that can resolve the issue by design should be used, whenever practical. Design solutions represent the first line of defense with the least uncertainty and will remain in effect over the life of the plant.

---

[22]Although mitigation is a safety function, security provides conditions that allow mitigation to occur during security related events.

Summary

The proposed security performance standards can be used to (1) describe the protection against security-related threats desired in new plants and (2) assess whether changes should be made to plant design, operation, preparedness or security to provide high assurance of protection of public health and safety, the environment, and the common defense and security.

The security performance standards provide specific criteria to implement the security expectations. They focus on protection of public health and safety and build upon and are consistent with existing safety philosophy (e.g., defense-in-depth) and expectations (e.g., safety goals), thus helping to integrate safety, security, and preparedness. They are risk-informed and written in a performance-oriented fashion expressed either in a qualitative fashion or a quantitative fashion. The proposed security performance standards also represent a policy issue needing the Commission's review and direction. Appendix C describes this issue further. The way in which these standards are used in decision making will require an integrated decision-making process, which is described in the next section.

6.7.3   Integrated Decision-Making Process

The results of the safety and security assessment should be reviewed for the security performance standards and other considerations important to deciding on the adequacy of security. This would require an integrated decision-making process that takes into account several factors that could influence the decision. These factors and their relevance to the decision are discussed below.

Security Performance Standards

The security performance standards listed above provide factors to be considered in the decision that can be related to performance of the plant. This performance can be calculated quantitatively or qualitatively. Each of these factors should be considered in the decision as follows:

- The guidelines of Figure 6-7 should be met for each threat considered, for both early and latent fatality risk. The cost benefit associated with region 2 of Figure 6-7 should be done in accordance with NUREG/BR-0058 "Revision 4, "Regulatory Analysis Guidelines of the USNRC." [NRC XXXX]

- The defense-in-depth provisions (i.e., multiple layers of defense and prevention and mitigation) should be met for each threat considered.

- Theft or diversion should be prevented by implementing the requirements in 10 CFR 73 with high assurance.

- Engineered design solutions to security-related issues are preferred, particularly when they reduce the need for human actions.

Other Factors

Other factors not included in the performance standards also should be considered in the decision. These factors are:

- The requirements in 10 CFR 73 and the post 9/11 orders should be complied with, unless an exemption is obtained.

- The scope and quality of the analysis used in the assessment should be consistent with the scope of the threat being assessed and with accepted methods and data.

- The impact of security-related actions (e.g., design changes, operational changes) should not detract from overall plant safety, preparedness or worker safety.

- Unquantified uncertainties should be considered for whether they could have a major influence on the decision.

The decision process used in evaluating whether to make a change in plant design, operation, or security as a result of the safety and security assessment would need to consider all of the above factors. Considering all of these factors will help to ensure integration of safety, security, and preparedness. Ideally, all factors should be met before deciding to take an action. However, this may not always be possible, in which case the factors for and against a change should be weighed, and the decision justified on a relative basis.

Finally, over the lifetime of the plant design, operational and security changes are likely to be proposed. Such changes should be evaluated for their impact on security using the same security performance standards and integrated decision-making process described above.

6.8   References

[ICRP 1984]   International Commission on Radiation Protection, ICRP 41: Non-stochastic Effects in Ionizing Radiation, 1984.

[ICRP 1993]   International Commission on Radiation Protection, ICRP Publication 64: Protection from Potential Exposure: A Conceptual Framework, Annals of the ICRP Volume 23/1, Elsevier, May 1993.

[EPA 1992]   U. S. Environmental Protection Agency, Office of Radiation and Indoor Air Radiation Protection Division, Protective Action Guides (PAG) Manual, 1992.

[NRC 1986]   U. S. Nuclear Regulatory Commission, "Safety Goals for the Operation of Nuclear Power Plants; Policy Statement," Federal Register, Vol. 51, p. 30028, August 21, 1986.

[NRC 1989]   U. S. Nuclear Regulatory Commission, "Health Effects Models for Nuclear Power Plant Consequence Analysis: Low LET Radiation," NUREG/CR-4214, 1989.

[NRC 1994]   Commission Policy Statement on the Regulation of Advanced Nuclear Power Plants, 59 FR 35461, 1994.

[NRC 1996]   U.S. Nuclear Regulatory Commission, "Standard Review Plan for the Review of Safety Analysis Reports for Nuclear Power Plants," NUREG-0800, Draft Rev. 3, June 1996.

[NRC 1997]   U.S. Nuclear Regulatory Commission, "Abnormal Occurrence Reports: Implementation of Section 208 Energy Reorganization Act of 1974," 62 FR 18820, 1997.

[NRC 1998] U.S. Nuclear Regulatory Commission, "Code Manual for MACCS2," NUREG/CR-6613, 1998.

[NRC 2004] U.S. Nuclear Regulatory Commission, "Regulatory Analysis Guidelines of the U.S. Nuclear Regulatory Commission," NUREG/BR-0058, Rev. 4, September 2004.

[NRC 2005a] U.S. Nuclear Regulatory Commission, "Security Design Expectations for New Reactor Licensing Activities," SECY-05-0120, July 6, 2005.

[NRC 2005b] U.S. Nuclear Regulatory Commission, "Staff Requirements - SECY-05-0120 - Security Design Expectations for New Reactor Licensing Activities," September 9, 2005.

[NRC 2007] U.S. Nuclear Regulatory Commission, "Revision of Policy Statement on Regulation of Advanced Reactors," SECY-07-0167, September 25, 2007.

7. LICENSING BASIS: PRA TECHNICAL ACCEPTABILITY

7.1 Introduction

The purpose of this chapter is to define the technical acceptability of the PRA needed to support the probabilistic process identified in this NUREG. The technical acceptability was defined in recognition of the increased role that PRA will play in the establishment of the licensing Framework for future reactors and the limitations of the current guidance, requirements and standards due to their specificity towards the existing light water reactors (LWRs). This subelement of the Framework is shown in Figure 7-1.

As stated in Chapter 1, it is expected that future applicants will rely on PRAs as an integral part of their license applications. This integration of PRA into the design and licensing process creates new challenges in the construction and maintenance of PRAs, and causes completeness, defensibility and transparency to be more important than it ever was in the past. Traditionally, the scope of LWR PRAs has been confined to the analysis of beyond design basis accidents, i.e., accidents leading to severe core damage. With the approach described in this document, the PRA and therefore the scope of the PRA needs to encompass a whole spectrum of off-normal events including frequent, infrequent and rare initiating events and event sequences. These events include a spectrum of releases from minor to major, and sequences that address conditions less than the core damage sequences of the current reactors and those similar to current reactor core damage sequences. It also needs to address the dose consequences of these event sequences as measured at the exclusion area boundary (EAB).

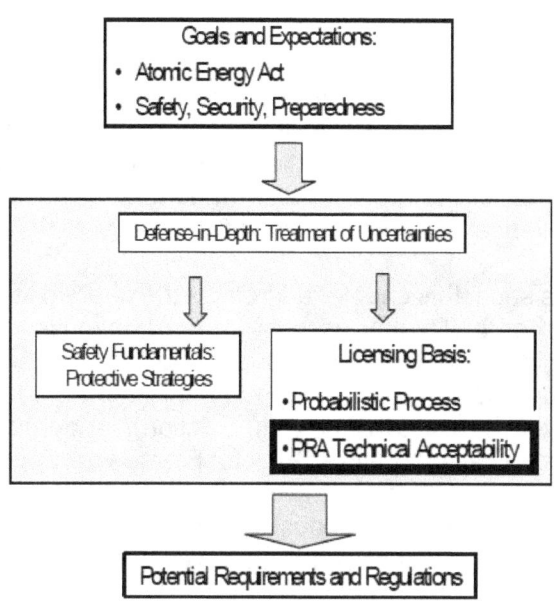

Figure 7-1   Framework Licensing Basis: PRA Technical Acceptability

The scope needed for the Framework is also broader than that typically considered in today's PRAs. It not only needs to address at-power and shutdown reactor operation, but also needs to be able to support the assessment of non-traditional events, such as, security.

One of the objectives of this document is to develop a basis for a regulatory structure that is applicable to all types of reactor designs, including gas-cooled, liquid metal, and heavy and light-water-moderated reactors. Current guidance, requirements and standards are constructed on the bases of applying PRAs to LWR applications. Metrics such as core damage and large early release may not be applicable to some advanced reactor designs. The current set of PRA levels that addresses progression to core damage, containment response and public-health consequences may also be less applicable as these are technology specific to LWRs. Therefore, in addition to issues associated with the role of the PRA, the applicability of the available guidance needs to be assessed and updated to reflect the application of PRAs in the Framework.

The consideration of uncertainties is a vital part of understanding the extent of the risk. Uncertainties need to be addressed in the calculation of both frequencies and consequences of the event sequences and the understanding of uncertainties is necessary for the evaluation of these event sequences against the potential requirements of the frequency-consequence (F-C) curve. Therefore, part of the examination of the design is identification, evaluation, and management of uncertainties.

Future reactor designs are likely to make more extensive use of passive systems and inherent physical characteristics to ensure safety, rather than relying on the active electrical and mechanical systems. As a result, the assessment of potential errors that occur during the design, manufacturing, fabrication and/or construction processes will be critical to ensuring safety. These latent errors are especially important for advanced designs, in which there is likely to be greater reliance than in the past on factory fabrication (as opposed to field fabrication).

As such, this chapter addresses the needed technical acceptability of the PRA scope and technical acceptability that account for the above issues. Section 7.2 provides a brief introduction to the selection of PRA as an analytic tool to support the implementation of this document. Section 7.3 addresses the application of the PRA in the Framework. It collects and summarizes the uses of the PRA in the Framework. Section 7.4 identifies the attributes necessary to ensure the scope and technical adequacy of PRA for Framework applications. This section briefly discusses the technical attributes of a quality PRA that are identified in Appendix F. The technical attributes are based on the existing PRA quality technical characteristics and attributes delineated in Regulatory Guide 1.200 and the High Level Requirements (HLRs) identified in the currently available PRA standards.

Additional methods to help establish PRA quality are also provided in Section 7.4. The methods include the establishment of a PRA quality assurance program, the use of consensus standards to delineate the requirements necessary to address the attributes identified in Appendix F, and an independent peer review process. The use of PRAs in the licensing of advanced reactors and operation (e.g., the maintenance of licensing-basis events (LBEs)) will require PRAs to be maintained current with the as-built, as-operated plant. Section 7.4.9 addresses the need to update and manage the configuration of the PRA to reflect changes in plant operation and design, and to review if past licensing decisions remain valid.

7.2   Probabilistic Risk Assessment

A PRA is an analytical technique for addressing risk as it relates to the performance of a complex system to understand likely outcomes, sensitivities, areas of importance, system interactions, and areas of uncertainty. It answers three questions: (1) What can go wrong? (2) How likely is it? (3) What are the consequences? These questions are often referred to as the "risk triplet."

The current body of regulations, guidance and license conditions is based largely on a "deterministic" approach. The deterministic approach establishes requirements for engineering margin and for quality assurance in design, manufacture, and construction. In addition, it assumes that adverse conditions can exist and establishes a specific set of design basis events (i.e., what can go wrong?). The deterministic approach involves imposed, but unquantified, elements of probability in the selection of the specific accidents to be analyzed as design basis events. It requires that the design include safety systems capable of preventing and/or mitigating the consequences (i.e., what are the consequences?) of those design basis events in order to protect public health and safety. Thus, a deterministic approach explicitly addresses only two questions of the risk triplet.

A probabilistic approach to regulation considers risk (i.e., all three questions) in a more coherent, explicit, and quantitative manner. It explicitly addresses a broad spectrum of initiating events and their event frequencies. It's structured approach includes each cause and effect that can result from a given initiating event. It includes an assessment of the impacts of degraded system, structure, component and human performance on the integrated plant response to the initiating events. It assesses the impact of multiple and common cause failures, and the complex interactions that can result from these failures. It also includes the impact of component and system unavailability due to maintenance or testing. The probabilistic approach enables the generation of event scenarios that consider the frequency and consequences of each initiating event and the associated response features (including human response) of the plant. It enables the analyses of the consequences of these event scenarios and weights the consequences by the frequency, thus giving a measure of risk. It allows for the determination of the functional importance of mitigation features and the explicit treatment of uncertainty. The probabilistic approach also allows for greater realism and minimizes the need for bounding analyses and the use of conservative margins.

While the traditional approach to regulation has been successful in ensuring no undue risk to public health and safety, opportunities for improvement exist. Given the broad spectrum of equipment and activities covered, the regulations can be strengthened and resources allocated to ensure that they are focused on the most risk-significant equipment and activities, and to ensure a consistent and coherent framework for regulatory decision-making [NRC 1998]. This NUREG uses the strength of the PRA's analytical techniques for addressing risk as a key element in the regulatory structure of future reactors.

7.3   PRA Applications in the Framework

PRA plays a significant role in the Framework. Its primary mission is to generate a complete set of accident sequences including a rigorous accounting of uncertainties. These sequences are used to evaluate the level of safety by comparing the PRA results with the Quantitative Health Objectives (QHOs) and the F-C curve. They are also used to generate the set of LBEs. These LBEs are assessed against the Framework's LBE acceptance criteria using the calculated frequency and consequence of the PRA sequences. The generation of sequences and LBEs, and other applications of the PRA within this document are presented using the following life-cycle phases of the plant: design, construction, startup and operation. High level PRA attributes for each phase are summarized below. The identified application bullets are explained in the sections following this summary.

Design

The PRA developed during the design phase will likely evolve as the design matures. Although the Framework is structured around an evolving PRA analysis that both influences the design and is influenced by the design, the PRA applications listed below are expected to be completed when the design is submitted for licensing.

- Generation of a complete set of accident sequences

- Development of a rigorous accounting of uncertainties

- Evaluation of the PRA results against the QHOs

- Evaluation of the PRA results against the F-C curve

- PRA supported assessment of security

- Identification and characterization of the LBEs

- Identification and characterization of the special treatment structures, systems and components (SSCs)

- Support the development of the environmental impact statement (EIS) and the severe accident mitigation design alternative (SAMDA) analysis

The PRA submitted for licensing is to reflect the proposed design and the expected operation and performance of the plant staff and equipment.

Construction

Future reactor designs will likely use passive systems and inherent physical characteristics to ensure safety, reducing the reliance on the active electrical and mechanical systems. However, fabrication and construction errors are one way in which design assumptions can be invalidated. Therefore, the identification of adverse latent conditions that could occur during fabrication and construction will be critical to ensuring safety. As such, risk-informed inspection insights will help focus inspectors to maximize the likelihood of identifying these conditions. In addition, changes that occur during construction need to be reflected back into the PRA and assessed for their impact on the level of safety. The following PRA-related activities are expected to occur during fabrication and construction.

- Maintain PRA
- Perform risk-informed inspections

Startup

This phase focuses on the initial staffing, training and programmatic issues that are expected to be finalized prior to startup and includes the following activities:

- Maintain PRA
- Support the determination of staffing requirements
- Support the development of the technical specifications (or equivalent)
- Support the development of inspection, testing and preventative maintenance
- Support the development of procedures and training
- Support the development of emergency preparedness (EP)

Operation

On completion of the design, construction and startup phases, the updated PRA reflecting the final design and operating philosophy will continue to be used to support licensing activities and plant operations. As this PRA is directly integrated into the design and licensing processes, it requires a comprehensive maintenance and update process. In addition, it is expected that a risk-informed philosophy will be integrated into the operation of the plant at a greater level than that of the current plants. The following activities are expected:

- Maintain PRA
- Assess and manage operational risk
- Assess and manage plant changes
- Monitor SSC performance
- Maintain a risk-informed training program

7.3.1   Generate a Complete Set of Accident Sequences

A key mission of the PRA analysis is to generate a complete set of accident sequences. These sequences are the foundation for many of the PRA's Framework applications and are a direct input into the determination of the proposed design's level of safety. They encompass a whole spectrum of off-normal events including frequent, infrequent and rare initiating events and event sequences. They include a spectrum of releases from minor to major, and sequences that address conditions less than the core damage sequences of the current reactors and those similar to current reactor core damage sequences. These sequences could also be used to aid in the application of the design's deterministic requirements including the assessment of barrier integrity requirements; protective system redundancy, diversity, reliability and availability requirements; and protective action effectiveness. As stated in Chapter 8, events that could defeat the protective systems, barrier integrity and protective action strategies simultaneously need to be identified and are required to be less than 1E-7 per plant year. The application of the PRA to each protective action strategy is discussed below.

Physical Protection

The PRA application to this protective strategy is discussed in Section 7.3.5.

Stable Operation

The PRA will be used to develop a complete set of initiating events. As discussed in Chapter 8, one of the potential defense-in-depth requirements will be to establish cumulative limits on frequency of these initiating events. Similar to the three categories developed for event sequences in Chapter 6, initiator events can also be divided into categories of frequent, infrequent and rare events each with a cumulative frequency limit. These limits will help to ensure there is a reasonable balance between plant challenges, accident prevention and mitigation.

Note that initiating event consideration for future designs may differ substantially from what is done for current LWRs. Given the unfamiliarity and lack of operating experience with advanced designs, search techniques such as master logic diagrams may have to be employed to identify initiators. This is similar to what was done early on in the application of PRA to LWRs, and has been done in the application of PRA to DOE facilities, medical systems, etc.

Protective Systems

Deterministic requirements have been established for the functions of reactivity control (reactor shutdown) and decay heat removal. A review of the PRA event sequences will aid in ensuring that for frequent and infrequent event sequences, there are redundant, diverse and independent means for reactor shutdown and decay heat removal as discussed in Chapter 8.

This NUREG utilizes reliability and availability goals for the SSCs within the PRA. Therefore, the PRA sequences should reflect these goals. These goals may vary depending on the sequence frequency. Protective systems responding to events that are expected to occur one or more times during the life of the plant (frequent events in Chapter 6) should have high availability and reliability, whereas protective systems that are in the design to respond to events not expected to occur (infrequent and rare events in Chapter 6) may have a lower availability and reliability. The results of the sequences will confirm the adequacy of these goals. Note that these goals need to be consistent with the expected performance of the equipment and will be monitored during the operation phase.

Protection against common-cause failures has been, and will continue to be important as these types of failures can dominate the unreliability of systems with some degree of built-in redundancy. The PRA provides a means of assessing the importance of common cause failures and provides the designer the ability to ameliorate the potential for these failures through selection of diverse materials, components, and manufacturing processes. It is worth noting that the current treatment of common cause failures is often data-driven, i.e., historical data is use to determine which common cause events are most likely and, hence, should be incorporated into the PRA. While some of this data may be relevant to future reactors, other information (including qualitative and quantitative screening) may be needed to identify significant common cause events associated with new or novel equipment.

Barrier Integrity

The PRA will aid in the determination of what barriers need to be in the design and how they should be designed. As discussed in Chapter 6, sequences that are categorized as frequent cannot contain any failed barriers and those categorized as infrequent needs to maintain containment functional capability. The PRA generated sequences will be used to verify that this requirement is met.

Barriers also need to be designed to maintain their integrity during the normal operational conditions such that their failure does not become an initiating event. The assessment of initiating event frequencies including those resulting from barrier failures such as loss of coolant accidents is in the scope of the PRA.

Protective Actions

The human action analysis used to support the development of the accident sequences needs to be consistent with the protective actions in the proposed emergency operating procedures (EOPs), accident management and EP procedures; and the proposed staffing levels. It is also expected that the EOPs, accident management and EP procedures will be developed with insights from the PRA such that all relevant accident PRA sequences are addressed. The analysis of accident sequences will also help to ensure that dependence on a single protective action included in these procedures does not prevent an exceeding of the F-C curve.

7.3.2   Develop a Rigorous Accounting of Uncertainties

In applying PRA to future reactor designs, analysts need to start with a clean page, i.e., not be biased by expectations from the conclusions of PRAs on old designs. Part of the examination of the unexpected is identification, evaluation, and management of uncertainties.

Uncertainties need to be addressed in the calculation of both frequencies and consequences of the event sequences. Since the sequences include rare events and event combinations postulated to occur in complex systems for which there may be limited experience, the consideration of uncertainties is a vital part of understanding the extent of the risk.

Uncertainties in some functional areas may make it difficult to conclude that adequate protection is provided and could lead to the need for additional safety enhancements, such as additional design features to provide more defense-in-depth, additional testing, and additional oversight by the NRC, all with the aim of achieving a high level of safety and confidence. Expected areas where high uncertainty related to modeling and completeness may be present include accident phenomenology, digital electronic instrumentation and control systems (including software and "smart" systems), human reliability, and passive system performance.

Sensitivity studies (e.g., alternative success criteria) are an important means for examining the impacts of modeling uncertainties. This will be of special use early in the licensing process, as the staff can use the PRA to highlight important areas of uncertainty where more research may be required to reduce the uncertainty, or, if the uncertainty cannot be reduced, where more defense-in-depth may be needed. The PRA can be used to examine the tradeoff between reducing the uncertainty through research and adding defense-in-depth or additional safety margin to cope with the uncertainty.

A range of uncertainties in future reactor performance needs to be considered including the following:

- Parameter uncertainty associated with the basic data; while there are random effects from the data, the most significant uncertainty is epistemic - is this the appropriate parameter data for the situation being modeled.

- Model uncertainty associated with analytical physical models and success criteria n the PRA can appear because of modeling choices, but will be driven by the state-of-knowledge about the new designs and the interactions of human operators and maintenance personnel with these systems.

- Completeness uncertainty associated with factors that are not accounted for in the PRA by choice or limitations in knowledge, such as unknown or unanticipated failure mechanisms, unanticipated physical and chemical interaction among system materials, and, for PRAs performed during the design and construction stages, all the factors affecting operations (e.g., safety culture, safety and operations management, training and procedures, use of new I&C systems).

All identified and quantified uncertainties (aleatory and epistemic) need to be included in the PRA that support the PRA applications within the Framework. The PRA directly uses the results of parameter estimation in the data uncertainty distribution for its basic events. It also uses many results of sensitivity studies to address uncertainty in success criteria, plant conditions and other models - sometimes incorporating model uncertainty, sometimes bounding it. It is important to qualitatively describe and catalog all aspects of uncertainty, even those difficult to quantify, for consideration in balancing structuralist and rationalist aspects of the Framework.

7.3.3 Evaluate the Quantitative Health Objectives (QHOs)

As stated in Section 3.1.2, the level of safety that future reactors are expected to meet are the QHOs for each event sequence and for the aggregate of all the event sequences. This means the PRA results need to demonstrate that the total integrated risk from the PRA sequences satisfies both the latent cancer QHO and the early fatality QHO. Therefore, the PRA sequences need to be able to characterize the offsite consequences of an accidental release of radioactive material in impacts on human health.

7.3.4 Evaluate the Frequency-Consequence Curve (F-C Curve)

As discussed in Chapter 6, the F-C curve relates the frequency of potential accidents to acceptable radiation doses at the EAB. The sequences of the PRA will populate the space under the F-C curve. Some scenarios will have little or no consequences, primarily because of the inherent characteristics and design features of the plant. Others are likely to approach the F-C curve and thus make up the important contributors to the plant risk profile. Therefore, in addition to the human health effects discussed in Section 7.3.3, the PRA event sequences need to be able to generate dose estimates. For frequent events, this estimate is for the annual dose to a receptor at the EAB. For infrequent and rare events, this estimate is for the worst two-hour dose at the EAB.

7.3.5 Support the Assessment of Security

It is expected that each future reactor will be required to perform a security assessment integral with the design and that PRA techniques could be used to identify combinations of equipment functions and operator actions that, if failed, could generate radiological releases. The scope and guidelines for performing the security assessment are discussed in Section 6.7.

7.3.6 Identify and Characterize the Licensing Bases Events (LBEs)

The identification and characterization of LBEs is derived from the PRA's accident sequences.

Identification

As discussed in Chapter 6, LBEs are bounding PRA event sequences that are used to provide assurance that the design meets the design criteria for various accident challenges with adequate defense-in-depth (including safety margin) to account for uncertainties. LBEs are chosen by grouping similar event sequences and associating an LBE with each grouping as described in Chapter 6.

Acceptance Criteria

In addition to meeting the F-C curve, LBEs categorized as frequent and infrequent are also required to meet deterministic requirements as described in Chapter 6. The PRA is used to identify event sequences that fall into various categories defined by frequencies that are ultimately identified as LBEs and assigned deterministic requirements based on these categories.

7.3.7 Identify and Characterize the Treatment of Safety-Significant SSCs

The PRA is used to both identify and characterize the safety-significant SSCs. Potential requirements for identification and treatment are discussed below.

Identification

The Framework's approach for selecting safety-significant SSCs is based on those SSCs that are relied upon to remain functional during the LBEs. Since the safety significant SSCs are linked to the LBEs and the LBEs were chosen in a risk-informed manner, the Framework approach for selecting SSCs for special treatment is also risk-informed. Other SSCs, besides those required for the LBEs may also be included based on risk importance measures that result from a plant-specific PRA.

Special Treatment

For those SSCs classified as safety significant, the special treatment they receive will vary since the treatment will be aligned with the mission the SSC needs to fulfill. In other words, the treatment ensures that the SSC will perform reliably (as postulated in the PRA) under the conditions (temperature, pressure, radiation, etc) assumed to prevail in the accident scenarios for which the SSC's successful function is claimed in the risk analysis.

7.3.8 Support the Environmental Statement and Severe Accident Mitigation Development

Section 102 of the National Environmental Policy Act (NEPA) (42 USC 4321) directs that an environmental impact statement (EIS) is required for major Federal actions that significantly affect the quality of the human environment. Included in the EIS is a requirement to assess alternatives to the proposed action. This requirement has been codified in 10 CFR 51, "Environmental Protection Regulations for Domestic Licensing and Regulatory Functions," as requiring an environmental impact statement for issuance of a permit to construct a nuclear power reactor, testing facility or fuel reprocessing plant. The environmental report is required to include an analysis that considers and balances the environmental effects of the proposed action, the environmental impacts of alternatives to the proposed action, and alternatives available for reducing or avoiding adverse environmental effects. This is commonly referred to as the Severe Accident Mitigation Design Alternative (SAMDA) analysis. The SAMDA analysis presents the environmental impacts of the proposal and the alternatives in comparative form. Where important to the comparative evaluation of alternatives, appropriate mitigating measures of the alternative need to be discussed. The PRA is used to support the identification and assessment of these alternatives.

7.3.9 Maintain PRA

The PRA used to support licensing needs to be maintained throughout the construction, startup and operation phases. During these phases, it is expected that the PRA will be maintained consistent with the plant's current performance and design. This will require the monitoring of SSCs included in the PRA to ensure their reliability, availability and performance are sufficient to support the goals of the design certification PRA. It will also require the monitoring of changes in the initiating event scope and frequency, modeling, software, industry experience, etc. In addition to monitoring the PRA inputs, a process will be required that evaluates the impact of deviations in performance or design, and maintains the risk-informed Framework applications and ultimately the level of safety.

The PRA will be maintained to a much greater extent than has been common practice for current LWRs. As such, and because the PRA is being used as an input to the plant's licensing basis, it is possible that changes in the PRA will result in the identification of new LBEs or safety significant SSCs as time passes or result in the shifting of an LBE from one frequency category to another.

The potentially dynamic nature of the identification and characterization of LBEs and safety significant SSCs makes a formal configuration change process a necessity.

Additional details on the configuration control program used to maintain the PRA is described in Section 7.4.9.

7.3.10   Risk-informed Inspections during Fabrication and Construction

The PRA will provide insights regarding the importance of various plant features and can be used to help identify items for inspection.

Construction errors are one way in which design assumptions can be invalidated. Techniques such as HAZOP may provide useful search schemes for identifying those construction errors that can cause the facility to operate outside the design assumptions. This will require creativity on the part of the PRA analysts beyond the routine that has arisen from the repeated application of PRA to the current generation of LWRs.

The PRA identification of safety-significant SSCs will provide a list of components for which it may be important for the NRC to conduct inspections during the fabrication process. This is especially important for advanced designs, where there is likely to be greater reliance than in the past on factory fabrication (as opposed to field fabrication). In addition, components may be fabricated outside the U.S., possibly to non-U.S. codes and standards.

7.3.11 Startup

Startup addresses those risk-informed activities that will likely continue to evolve following the receipt of a license but need to be in place prior to reactor startup. These are discussed below:

Staffing Evaluation

The burden is on the applicant to demonstrate through modeling of human actions, the use of simulators and/or mockups, and the PRA analysis that staffing is adequate for the evaluated level of safety. The determination includes the assessment of human actions needed which should be consistent with those in the PRA and the reliability of these actions assumed in the PRA. Consideration needs to be given to conditions that could shape the action's failure probability such as: its complexity, time available for action completion, procedure quality, training and experience, instrumentation and controls, human-machine interface and the environment.

Technical Specifications

Technical specifications of the past prescribed out-of-service equipment configurations with specific allowed outage times (AOTs) and action statements. In contrast, advanced designs may rely much more on risk-informed technical specifications, where allowed equipment configurations and AOTs are fluid, changing as the plant configuration changes. The risk impact of configuration changes will likely be measured by a risk monitor, which in turn relies on the plant PRA. Furthermore, the PRA (input to a risk monitor) will consider all modes of plant operation and will consider both internal and external initiators. This treatment of the plant configuration will be a more integrated assessment than that for current LWRs.

Lessons learned from efforts to risk-inform the technical specifications for currently operating LWRs should be considered in developing the technical specification requirements and any implementing guidance.

Inspection, Testing and Preventive Maintenance

With regard to inspection and testing, the requirements need to be set consistent with the importance of a particular SSC within the PRA. Preventive maintenance designed to maintain an assumed reliability should be balanced against increased unavailability of the SSC resulting from the preventive maintenance. It is envisioned that the SSC reliability monitoring program discussed in Section 7.3.9, Maintain PRA, will also support the process of developing an effective maintenance program.

Development of Procedures and Training

The PRA can provide valuable insights regarding the importance of human actions, which can then be emphasized in procedures and training programs. It is expected that procedural guidance will be developed for all actions credited within the PRA and that training will be risk-informed.

Development of Emergency Preparedness

The analysis of the plant PRA helps to determine the measures that are effective in limiting the public health effects from radionuclide release accidents so that public health effects remain below the limits set in the QHOs.

7.3.12 Operation

The integration of PRA into the design and licensing process enables a coherent application of risk into operational processes associated with the design, operation and maintenance of the plant. Some of these processes are discussed below.

Assess and Manage Operational Risk

The Maintenance Rule (10 CFR §50.65) establishes the requirements for monitoring the effectiveness of maintenance at current plants. It is envisioned that a similar program will exist for future reactors including assessment and management of the increase in risk that may result from proposed maintenance activities. For current reactors, this assessment is to be performed prior to the implementation of the maintenance activities.

Assess and Manage Plant Changes

In 10 CFR §50.59, "Changes, Tests and Experiments," the process for which a current licensee may make changes in the facility or procedures as described in the final safety analysis report and conduct tests or experiments not described in the final safety analysis report without obtaining a license amendment is addressed (10 CFR §50.59). As the licensing bases of future reactors are risk-informed, it is expected that the future reactor change process will be fully integrated with the Framework's LBE risk-informed acceptance criteria and deterministic defense-in-depth requirements.

Monitor SSC Performance

The SSC monitoring program confirms the reliability, availability and performance of equipment assumed in the licensing process and is a part of the PRA configuration control program described in Section 7.4.9.

Maintain a Risk-informed Training Program

Insights gained from the PRA can help ensure safe operation and need to be integrated into the operation and technical support staff training programs. This will ensure the staff is knowledgeable of the potential initiating events and analyzed accident sequences.

7.4 Functional Attributes for PRAs for Future Plants

This section addresses the functional attributes for PRAs for future plants used to support the Framework. These attributes address the following topics:

- Technical Attributes
- Quality Assurance Criteria
- Consensus Standards
- Assumptions and Inputs
- Analytical Methods
- Analytical Tools
- Independent Peer Review
- Documentation
- Configuration Control

Each of these topics is discussed below.

7.4.1 Technical Attributes

Appendix F identifies the high level attributes necessary to ensure the technical adequacy of a PRA used in the licensing, construction, startup and operation of a future reactor. The required scope of the PRA and the corresponding attributes for each technical element are addressed. Specifically, high level attributes are provided for all the technical elements of a PRA required to calculate the frequency of accidents, the magnitude of radioactive material released, and the resulting consequences. Attributes are also provided for the scope of the PRA which is defined by identification of the complete set of challenges including both internal and external events during all modes of operation.

A key mission of the PRA analysis is to generate a complete set of accident sequences. These sequences are the foundation for many of the PRA's Framework applications and are a direct input into the determination of the proposed design's level of safety. They include a spectrum of releases from minor to major, and sequences that address conditions less than the core damage sequences of the current reactors and conditions similar to current reactor core damage sequences. This is illustrated by the event tree shown in Figure 7-2.

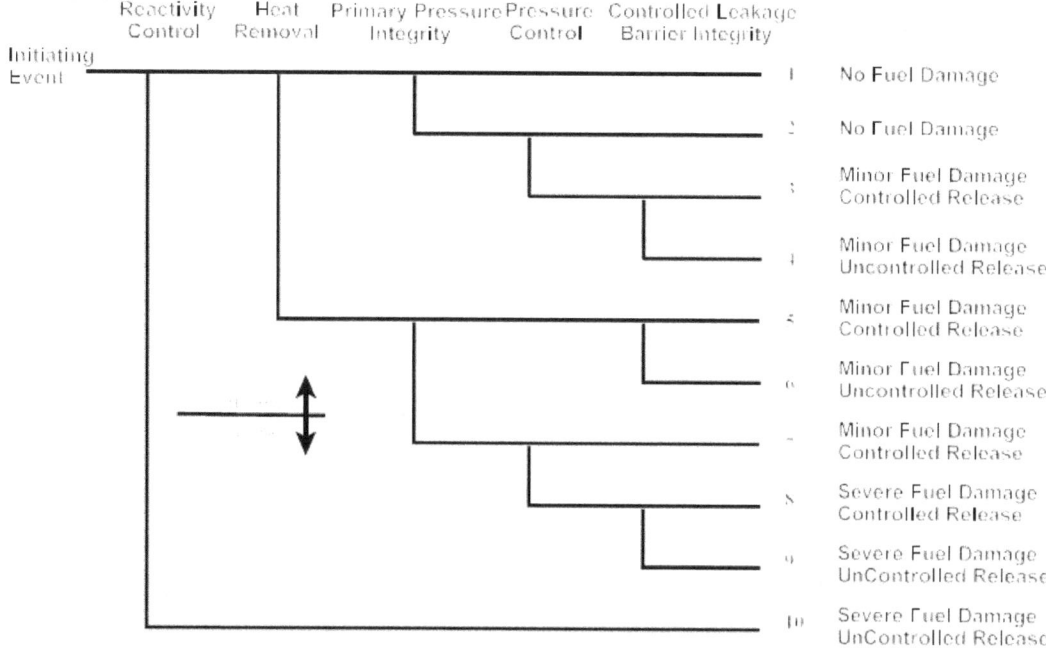

Figure 7-2   Event sequence example.

The functions and end states included in this event tree are for illustration purposes only and are not meant to represent a current or future reactor. For current PRAs, Sequences like 1 and 2 would be considered successful and therefore would not be included in the results of a current PRA. Sequences similar to 3 through 7 may or may not be included in current PRAs depending on the degree of fuel damage and the degree of conservatism used in the PRA. Sequences 8 through 10 would be core damage sequences included in the results. For a PRA supporting this document, all these sequences are of interest. Sequences 1 and 2 would likely be considered frequent events, depending on the initiating event frequency and the failure probabilities of the event tree functions, with no or limited release of radionuclides. Given no fuel damage, the potential for radionuclide release would be bounded by the allowable activity within the primary coolant or activity associated with other radiological sources. Sequences 3 through 7 would likely have higher allowable doses as a result of the limits established by the F-C curve discussed in Chapter 6 and the lower frequencies of these sequences due to the increased number of failures. These intermediate sequences may require that different levels of system success criteria be defined within the PRA models to properly assess the dose consequences. Sequences 8 through 10 would need to be of a low enough frequency to allow the higher doses that will likely result from severe fuel damage.

Initiating event consideration may also be substantially different from those for current U.S. LWRs. Examples are events associated with on-line refueling, recriticality due to more highly enriched fuel and fuels with higher burnup, and chemical interactions with some reactor coolants or structures. For these reasons, more emphasis will be required on the use of systematic methods to identify the initiating events modeled in the PRA. Searches for applicable events at similar plants, if available (both those that have occurred and those that have been postulated), and use of existing deductive methods (e.g., master logic diagrams, top logic models, fault trees, and failure modes and effects analysis) could both be utilized in this effort.

Appendix F builds on existing PRA technical characteristics and attributes delineated in Regulatory Guide 1.200 and the High Level Requirements (HLRs) identified in currently available PRA standards. The attributes focus on a PRA of the reactor core but also address other radioactive materials (e.g., spent fuel and radioactive waste) that need to be considered to effectively evaluate risk.

7.4.2 Quality Assurance Criteria

The PRA analyses supporting this NUREG need to be subject to quality control. Given the integration of the PRA into the design and licensing process, the need for completeness, defensibility and transparency are more important than ever in the past. This translates directly into the need for more rigorous quality control requirements than those that are typical for PRAs supporting current reactors. Table 7-1 lists high level quality control attributes for a PRA that is supporting a Framework analysis (10 CFR 50 Appendix B and ANS standard [ANS 1974)].

Table7-1　　PRA quality assurance.

Topic		Attribute
1	Quality Assurance Program	At the earliest practicable time, consistent with the schedule for developing, modifying and maintaining the PRA, a quality assurance program needs to be established with written policies, procedures, or instructions and needs to be carried out throughout the life cycle of the analysis.
2	PRA Staff	Measures are established to provide for indoctrination, training and qualification of personnel performing PRA-related activities to assure awareness in quality assurance processes and controls and to ensure suitable technical proficiency is achieved and maintained.
3	Requirements and Standards	Measures are established to assure that applicable regulatory requirements and standards are specified and included in the development and maintenance of the PRA and that deviations from such standards and requirements are controlled.
4	Interface Control	Measures are established for the identification and control of PRA process interfaces and for coordination among interfacing design organizations. These measures need to include the establishment of procedures among participating organizations for the review, approval, release, distribution and revision of documents.
5	Independent Reviews	The PRA control measures need to provide for verifying or checking the adequacy of the PRA, such as by the performance of independent checks and peer reviews. The independent verifying or checking process needs to be performed by individuals or groups other than those who performed the original analysis, but may be from the same organization. In addition to the independent checks, an independent peer review process, as described in Section 7.4.7, needs to be performed.
6	Procedures	Activities affecting PRA quality are prescribed by documented instructions or procedures and need to be accomplished in accordance with these instructions or procedures.
7	Document Control	Measures are established to control the issuance of PRA documents. These measures need to assure that documents, including changes, are reviewed for adequacy and approved for release by authorized personnel. Changes to documents need to be reviewed and approved by the same organizations that performed the original review and approval unless designated to another responsible organization.

Table 7-1   PRA quality assurance.

Topic		Attribute
8	Corrective Actions	Measures are established to assure that conditions adverse to PRA quality are promptly identified and corrected. In the case of significant conditions adverse to quality, the measures need to assure that the cause of the condition is determined and corrective action taken to preclude repetition. The identification of the significant condition adverse to quality, the cause of the condition, and the corrective action taken is documented and reported to appropriate levels of management.
9	Audits	A comprehensive system of planned and periodic audits is carried out to verify compliance with all aspects of the quality assurance program and to determine the effectiveness of the program. The audits are performed in accordance with written procedures or check lists by appropriately trained personnel not having direct responsibilities in the areas being audited. Audit results need to be documented and reviewed by management having responsibility in the area audited. Followup action, including reaudit of deficient areas, need to be taken where indicated. The need for audits is in addition to the performance of an independent peer review which is discussed in Section 7.4.7.

7.4.3   Consensus PRA Standards

One acceptable means of demonstrating conformance with the regulatory approach is to use an industry consensus PRA standard or standards that address the scope of the PRA used in the decision making. The PRA standard needs to be applicable to the design of the plant and to the PRA applications specified in this NUREG.

7.4.4   Assumptions and Inputs

7.4.4.1   Assumptions

Assumptions used in the PRA supporting a Framework analysis need to be realistic and defensible with their basis and application clearly documented. They should not use significantly conservative or optimistic assumptions and should not use expert judgement except in those situations in which there is a lack of available information regarding the condition or response within the PRA, or a lack of analytical methods upon which to base a prediction of a condition or response. The assumption also should not take credit for SSCs beyond rated or design capabilities or heroic human actions that have a small probability of success. The PRA needs to include an assessment of uncertainty of the results for important or key assumptions both individually and in logical combinations.

Key assumptions are those that are related to an issue in which there is no consensus approach or model (e.g., choice of data source, success criteria, reactor coolant pressure seal loss-of-coolant model, human reliability model) and in which the choice of approach or model has an impact on the PRA results in terms of introducing new accident sequences, changing the relative importance of sequences, or affecting the overall results.

7.4.4.2 Inputs

All inputs need to be traceable to a clearly identified source and consistent with the proposed design.

7.4.5 Analytical Methods

The analytical methods used in a PRA supporting a Framework analysis need to be sufficiently detailed as to its purpose, method, assumptions, design input, references and units such that a person technically qualified in the subject can review and understand the analysis and verify the adequacy of the results without recourse to the originator. Where possible, analytical methods need to be consistent with available codes and standards and checked for reasonableness and acceptability. Method-specific limitations and features that could impact the results should be identified.

7.4.6 Analytical Tools

PRA quantification software, thermal/hydraulic codes, structural codes, radionuclide transport codes, human reliability models, common cause models, etc. are typically used in the PRA quantification process. These models and codes need to have sufficient capability to model the conditions of interest and provide results representative of the plant and need to be used only within known limits of applicability. As errors in such programs may significantly impact the results, it is necessary that the development and application of the computer programs, spreadsheets or other calculation methods exhibit a high level of reliability as ensured through a documented verification and validation process. Verification is a systematic approach to ensure the model or computer code correctly represents the model or code's design. Validation is the demonstration that the verified models or codes meet the requirements. In addition, users need to demonstrate the appropriateness of the models or codes selected for a specific application and of the way in which these programs are combined and used to produce the needed results [ANS 1987].

7.4.7 Independent Peer Review

A PRA that supports a Framework application should be peer reviewed. An adequate peer review is one that is performed by qualified personnel, according to an established process that compares the PRA against a set of desired characteristics and attributes, documents the results, and identifies both strengths and weaknesses of the PRA.

7.4.7.1 Team Qualifications

Team qualifications determine the credibility and adequacy of the peer reviewers. To avoid any perception of a technical conflict of interest, the peer reviewers will not have performed any actual work on the PRA. Each member of the peer review team needs to have technical expertise in the PRA elements he or she reviews, including experience in the specific methods that are used to perform the PRA elements. This technical expertise includes experience in performing (not just reviewing) the work in the element assigned for review. Knowledge of the key features specific to the plant design and operation is essential. Finally, each member of the peer review team needs to be knowledgeable in the peer review process, including the desired characteristics and attributes used to assess the adequacy of the PRA.

7.4.7.2 Peer Review Process

The peer review process includes a documented procedure used to direct the team in evaluating the adequacy of a PRA. The review process compares the PRA against desired PRA characteristics and attributes. In addition to reviewing the methods used in the PRA, the peer review determines whether the application of those methods was done correctly. The PRA models are compared against the plant design and procedures to validate that they reflect the as-built and as-operated plant. Key assumptions are reviewed to determine if they are appropriate and to assess their impact on the PRA results. The PRA results are checked for fidelity with the model structure and for consistency with the results from PRAs for similar plants based on the peer reviewer's knowledge. Finally, the peer review process examines the procedures or guidelines in place for updating the PRA to reflect changes in plant design, operation, or experience.

7.4.8 PRA Documentation

A PRA used in a Framework application should be documented such that a person technically qualified in PRA can review and understand the analyses and verify the adequacy of the results without recourse to the originator. The documentation needs to be traceable and defensible with sources of information both referenced and retrievable. It needs to support the determination that the PRA is performed consistent with the applicable standards and the technical attributes contained within this document. The documentation also needs to be maintained current with the plant configuration and the PRA model. The methodology used to perform each aspect of the work needs to be described either through documenting the actual process or through reference to existing methodology documents. Key sources of uncertainty need to be identified and their impact on the results assessed. Key assumptions made in performing the analyses need to be identified and documented along with their justification to the extent that the context of the assumption is understood. The results (e.g., products and outcomes) from the various analyses need to be documented. This documentation entails both submittal and archival documentation. PRA information submitted in support of a future reactor application will form part of the licensing basis and, as such, is expected to be docketed.

7.4.8.1 Submittal Documentation

To demonstrate that the technical adequacy of the PRA used in an application is consistent with the expectations and high level attributes identified in this NUREG, the following information should be submitted to the NRC:

- **Quality Assurance** – a description of the PRA quality assurance program including the methods used to implement the attributes of Section 7.4.2. Also include a description of the PRA configuration control program used to implement the attributes of Section 7.4.9.

- **Scope and General Methodology** – a description of the scope of the PRA and the overall methodology used in the analysis.

- **Technical Attributes and Consensus Standards** – documentation that demonstrates that the PRA is performed consistent with this NUREG's technical attributes, identification of the consensus standards applied to address those attributes, and a description of the extent of their application.

- **Assumptions and Inputs** – a description of all assumptions, their bases and applications. For key assumptions or other sources of uncertainty, an assessment of the impact of the uncertainty on the results, both individually and in logical combinations needs to be included. These assessments provide information to the NRC staff in their determination of whether the use of these assumptions and approximations is appropriate and whether sensitivity studies performed to support the decision are appropriate. A list of significant inputs and their application also should be provided.

- **Analytical Methods** – a description of all analytical methods, including selection of empirical factors, data inputs and limitations.

- **Analytical Tools** – a description of all analytical tools including models and computer codes and the verification and validation process used to ensure their accuracy.

- **Logic Models** – all event trees and fault trees with supporting bases information. Include information on structure, initiating events, top events and basic events, including human actions and common cause.

- **Reliability and Availability Data** – a description of the approach used to develop the reliability and availability data including a discussion on the application of reliability and availability goals.

- **Results** – the necessary results to demonstrate that the acceptance criteria are met including the identification of key sources of uncertainty and the treatment of uncertainty within the analysis.

- **Peer Review** – a discussion of the peer review process, the results of the peer review for each reviewed element and comment resolution specifying the action taken to address and resolve identified issues. Also include descriptions of the peer review team members and their qualifications.

7.4.8.2 Archival Documentation

This documentation includes all supporting calculations, procedures and references that were used to demonstrate that the acceptance criteria are met. This documentation is to be maintained as lifetime quality records in accordance with Regulatory Guide 1.33, by the applicant, as part of the normal quality assurance program, so that it is available for examination.

7.4.9 Configuration Control

The PRA used to support this NUREG needs to be maintained throughout the construction, startup and operation phases of the plant. Therefore a PRA configuration control program should be developed early in the design process and needs to be in place at the time the PRA is submitted for NRC staff review. This program includes the following key elements:

- a process for monitoring PRA inputs and collecting new information.

- a process that maintains and upgrades the PRA to be consistent with the current configuration of the plant as it progresses through construction, startup and operation.

- a process that ensures that planned plant and procedure changes are assessed prior to their implementation to ensure that the licensing acceptance criteria and deterministic defense-in-depth requirements are maintained valid.

- a process that ensures that unplanned changes in performance, new insights or methods, previously unidentified deficiencies or other issues impacting the PRA results are assessed in a timely manner to ensure that the licensing acceptance criteria and deterministic defense-in-depth requirements are maintained valid.

- a process that ensures the cumulative impact of pending changes is considered when applying the PRA.

- a process that evaluates the impact of changes on any other previously implemented risk-informed decisions that have used the PRA.

- a process that maintains configuration control of computer codes used to support PRA quantification.

- a process that maintains the documentation of the Program and provides periodic reporting of updated results to the NRC.

These elements are based on the requirements for PRA configuration controls for current reactors [ASME 2005] with modifications reflecting the various phases of the plant's life cycle and the integral role PRA plays in the licensing process. The integral role of the PRA results in the need for a more integrated risk-informed monitoring and change evaluation process and specific reporting requirements.

During the construction, startup and operation phases, planned plant and/or procedure changes are to be evaluated for their impact on the licensing acceptance criteria and deterministic defense-in-depth requirements prior to their implementation. This process is expected to be similar to the current 10 CFR §50.59 process where a safety evaluation screening process is typically performed on all proposed changes. Proposed changes need to be consistent with this document's acceptance criteria prior to implementation of the proposed change.

During operation, a process similar to the monitoring of the performance and condition of SSCs, against licensee-established goals of the Maintenance Rule (10 CFR §50.65), is expected to be an integral part of the monitoring program for the PRA. This process will use this NUREG's reliability and availability goals as the key input for the operational phase monitoring program. Monitoring will consist of periodically gathering, trending and evaluating information pertinent to the performance, and/or availability of PRA-related SSCs and comparing the result with the established goals and performance criteria to verify that the goals are being met [NUMARC 2000]. When the goals are met, the plant's performance is consistent with the licensing bases. When a goal or performance criteria is not met, then assessment of the impact of the performance issue on the PRA and licensing bases is required. Cause determination and corrective actions may also be required. Performance issues that result in the failure of this NUREG's acceptance criteria will require licensing action.

Unexpected changes in performance, methods or knowledge can result in changes to the PRA and to the frequency or consequence of identified LBEs or in the identification of new LBEs not previously analyzed. These changes can also impact the identification of safety-significant SSCs.

This document encourages the use of design margin to minimize the impact of PRA changes on the licensing bases. Changes that reduce margin but do not impact the Framework's regulatory safety margin will not require a re-assessment of the LBEs or the defense-in-depth measures. For plants built according to a certified design, if a proposed change modifies the certified (Tier 1 or Tier 2) portion of the design, a rule change to amend the certification or an exemption request is required.

The PRA update frequency is primarily dependent on the scope and nature of pending changes. It is expected that frequent updates will be performed to minimize the need to separately assess pending changes and to avoid the potential complexity of evaluating the cumulative impact of these changes. A maximum update interval of 5 years is proposed. It is expected that a report containing all plant changes, tests and experiments consistent with similar 10 CFR §50.59 requirements, including a summary of the risk-informed evaluation of each change will be submitted at intervals similar to that of the 10 CFR §50.59 reporting requirement (report of plant changes, tests and experiments at a frequency not exceed 24 months). Included with these submittals is an assessment of the cumulative impact of the changes, including unplanned changes as described above, and an assessment of these changes on the plant's safety margin.

7.5   References

[ANS 1987]   "American National Standard Guidelines for the Verification and Validation of Scientific and Engineering Computer Programs for the Nuclear Industry." ANSI/ANS-10.4-1987. 1987.

[ANS 1974]   "Quality Assurance Requirements for the Design of Nuclear Power Plants." ANSI N45.2.11, 1974.

[ASME 2005]   "Standard for Probabilistic Risk Assessment for Nuclear Power Plant Applications." ASME RA-Sb-2005. December 30, 2005.

[NUMARC 2000]   "Industry Guideline for Monitoring the Effectiveness of Maintenance at Nuclear Power Plants." NUMARC 93-01, Revision 3. July 2000.

NRC 1998]   Paper on Risk-Informed and Performance-Based Regulation." SECY-98-144. NRC: Washington, D.C. June 22, 1998.

8. INTEGRATED PROCESS: REQUIREMENTS DEVELOPMENT

8.1 Introduction

The purpose of this chapter is to (1) describe the process for integrating the various elements to develop risk-informed and performance-based technical and administrative requirements, and (2) illustrate the results of application of the process. As such, this chapter addresses the questions raised in Chapter 2: How should the elements of the Framework be integrated? Should each of the Framework elements be treated and addressed separately and independently? Or, should they be integrated to achieve a cohesive set of requirements and regulations? This final element of the Framework is shown in Figure 8-1.

This process is for use by the U.S. Nuclear Regulatory Commission's (NRC's) staff as guidance in developing technical and administrative requirements. Appendix J provides a set of potential requirements developed using the process described in this chapter and the guidance in this NUREG.

Since a set of stand-alone requirements is envisioned, it needs to be complete with respect to the technical and administrative risk-informed and performance-based

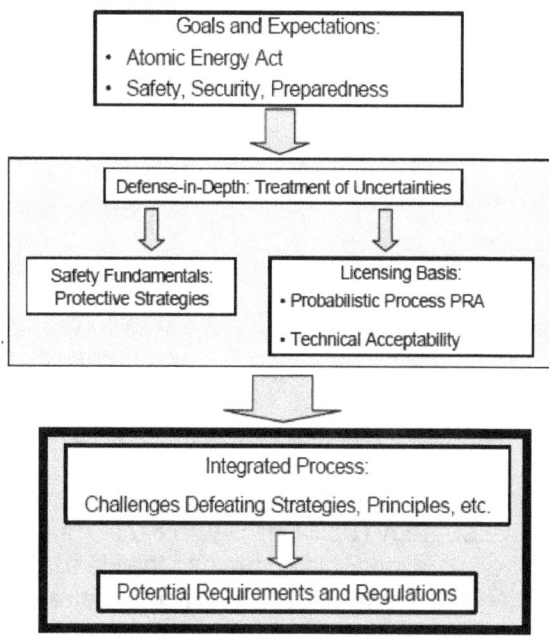

Figure 8-1   Framework Integrated Process

requirements, including interfacing with the other parts of 10 Code of Federal Regulations (CFR). Accordingly, since 10 CFR 50 has some requirements that are already applicable to all reactor technologies, it would make sense to carry these over into the new requirements since there is already implementing guidance and experience in their use. It is envisioned that this be accomplished by using as many of the existing 10 CFR 50 requirements and general design criteria as is reasonable, based upon their risk and performance-based characteristics. Appendix H of this NUREG identifies those 10 CFR 50 requirements and general design criteria recommended to be used. Where those 10 CFR 50 requirements and general design criteria also have Regulatory Guides (RGs), it is intended to use those same RGs for additional implementing guidance. Where new RGs are needed for the new technical requirements, they should be developed as needed. Appendix J also contains suggested RG content for the potential requirements.

8.2 Topic Identification Process

The Framework structure described in Chapters 2 through 7 defines an overall set of safety objectives, protective strategies, and criteria for an approach for future plant licensing applicable to diverse reactor technologies. The next step is to identify and define the topics where technical and administrative requirements are needed to ensure the safety objectives, protective strategies, and criteria in Chapters 2 through 7 are met. The process for the identification of the topics is shown in Figure 8-2. Each of the boxes shown in Figure 8-2 is discussed in the subsections below.

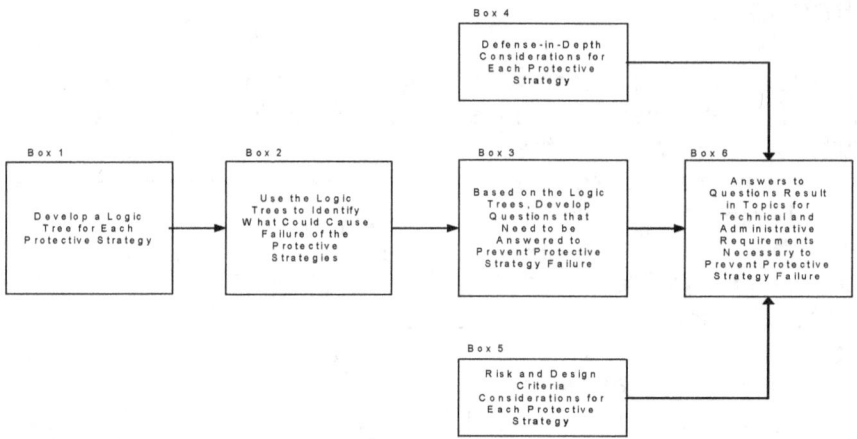

Figure 8-2   Process for Identification of
Requirement Topics

8.2.1   Logic Tree Development (Figure 8-2: Box 1)

For each protective strategy, a logic tree is developed that identifies what could cause failure of the protective strategy (Box 1 of Figure 8-2). As discussed in Chapter 5, these logic trees have been developed in a deductive manner that leads to the potential root cause of the failure, that is, identifying the different ways in which the strategy under consideration can fail. An example logic tree is shown in Figure 8-3. The actual logic trees for each protective strategy are contained in Appendix G. At the top level, each logic tree has three branches. These branches represent three basic pathways that can lead to the failure of a protective strategy.

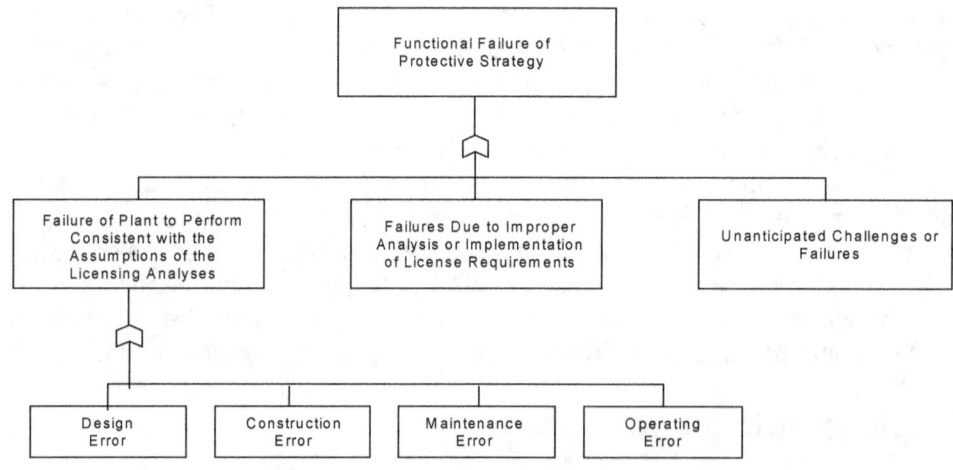

Figure 8-3   Example Logic Tree

All the protective strategy logic trees follow the same basic top logic structure to organize the kinds of failures that can occur and, therefore, to help identify the topics for which requirements are needed to develop confidence in the performance of the strategies.

The "Functional Failure of a Protective Strategy," as shown in Figure 8-3, can occur in one of the following three ways:

- failure of plant to perform consistent with the assumptions of the licensing analyses
- failures due to improper analysis or implementation of licensing requirements
- unanticipated challenges/failures

These three pathways are described below.

Failure of Plant to Perform Consistent with the Assumptions of the Licensing Analyses —

The licensing analysis (probabilistic risk assessment (PRA), deterministic, licensing-basis event (LBE) calculation, etc.) reflects the design expectations for protective strategy performance. Even though the licensing analysis may be correct, if the design is not implemented and maintained in a way to ensure the continuing validity of the licensing analysis, the protective strategy may not perform its anticipated function. For example:

- Errors in the detailed implementation of the design requirements (e.g., specific pump or valve selection or digital instrumentation and control (I&C) errors) introduce errors in system and component performance.

- Construction or installation errors that substitute improper equipment, introduces flaws, or that impede proper operation (e.g., inadequate ventilation of electronic equipment) also can introduce errors in system and component performance.

- Maintenance errors can disable equipment beyond the availability and reliability assumptions of the PRA (e.g., installing an improper software update can fail all redundant I&C and protection equipment), thus introducing errors in performance.

- Operations errors also can defeat redundancy and lead to failures beyond those modeled in the PRA. In particular, systematic change in procedures, training, or crew practices (e.g., communications, evaluation, use of procedures) can have sweeping effects.

- Updating the PRA (which is a key part of the licensing analysis) as the design becomes specific, after construction for the as-built plant, routinely as equipment performance and operations practices change, and as evidence of aging accumulates, can reduce the likelihood of failing to perform consistent with the licensing analysis.

Failures Due to Improper Analysis or Implementation of Licensing Requirement"—

The licensing analysis is the bases for believing that the design meets the NRC's safety, security and preparedness expectations. Confidence is also needed that the design is implemented, constructed, operated and maintained in accordance with the requirements. Failure to properly implement these requirements (due to errors in the analysis or interpretation of requirements) means that the risk could become greater than expected. Such errors can obscure or mask the risk and lead to unanalyzed events.

Unanticipated Challenges/Failures—

This pathway acknowledges completeness uncertainty. There may be initiating events and scenarios not identified in the PRA. While some systemic uncertainty always remains, it can be reduced in a number of ways:

- More thorough and systematic search schemes can be developed for identifying initiating events and scenarios in the PRA.

- Experimental and test programs can address technical knowledge gaps, both basic knowledge gaps and performance under unusual conditions.

- Application of the protective strategies and defense-in-depth provisions to help compensate for the uncertainty.

8.2.2   Protective Strategy Failure Causes (Figure 8-2: Box 2)

Under each of these three basic pathways, additional branches were developed to identify the root causes of failures that could lead to failure of the protective strategy. This was then used as a guide to identify questions that, when answered, would identify the topics that the requirements need to address to guard against the root cause of failure, consistent with the overall safety philosophy and criteria discussed in Chapters 2 through 7. The root causes of failure were identified using engineering judgement and previous experience with safety and PRA analysis and plant design, operation and experience.

8.2.3   Protective Strategy Failure Prevention (Figure 8-2: Box 3)

The end point of each branch developed in the logic trees translates into one or more questions corresponding to each of the potential root cause failures. That is, based on the causal events (or the basic events in the fault tree), a series of questions (Box 3 of Figure 8-2) were developed, the answers to which identify the actions that need to be taken to ensure the protective strategy is successful. These answers may be related to design, construction, maintenance, or operation. To facilitate going from the logic trees to the questions, each end point of each branch in the logic trees and each question corresponding to that end point have a unique identifying number (e.g., pp-1 for physical protection - question 1). Table 8-1 provides an example of the questions and answers for the Barrier Integrity Protective Strategy, organized by their applicability to design, construction or operation (in this document, maintenance is included in operation).

Table 8-1   **Example questions and answers - barrier integrity.**

Protective Strategy Questions	Answers to Questions		
	Design	Construction	Operation
Failure to Perform Consistent with Assumptions - Operations Error			
• What needs to be done to prevent operational errors?	• Use good human factors (HF) and human-machine interface engineering • Use fault tolerant designs	• N/A	• Verified procedures • Training • Use of simulator • Work control • Surveillance • In-service inspection (ISI) • In-service testing (IST)
Failures Due to Improper Analyses or Implementation of Requirements			
• How can failures due to improper analyses or implementation of requirements be prevented?	• Use verified analytical tools • Quality PRA and safety analyses • Quality Assurance (QA) (i.e., Ensure plant is designed consistent with PRA and safety analysis)	• QA/Quality Control (QC) (i.e., Ensure plant is constructed consistent with design)	• Technical Specifications • Monitoring and feedback
Failures Due to Challenges Beyond What Were Considered in the Design			
• How can challenges beyond what were considered in the design (i.e., uncertainties) be accounted for?	• At least 2 barriers for the prevention of releases of fission product (FP) from the reactor • Provisions to establish a containment functional capability independent of fuel and Reactor Coolant System (RCS) for the reactor	• N/A • N/A	• Emergency Preparedness (EP)

8.2.4   Defense-in-depth Considerations (Figure 8-2: Box 4)

In developing the answers to each question, other items also need to be considered (Boxes 4 and 5 of Figure 8-2). Box 4 represents the application of the defense-in-depth principles, discussed in Chapter 4, to each protective strategy to ensure that uncertainties (particularly completeness uncertainties) are properly considered at the protective strategy level. This can also result in additional topics being identified. For convenience, the defense-in-depth principles are repeated below:

- Measures against intentional as well as inadvertent events are provided.

- The design provides accident prevention and mitigation capability.

- Accomplishment of key safety functions is not to be dependent upon a single element of design, construction, maintenance, or operation.

- Uncertainties in structures, systems and components and human performance are accounted for in the safety analyses and appropriate safety margins are provided.

- The plant has alternative capabilities to prevent an unacceptable release of radioactive material to the public.

- Plants are sited at locations that facilitate protection of public health and safety.

The protective strategies also represent a high-level defense-in-depth structure for developing the requirements, in that each one represents a line of defense against the uncontrolled release of radioactive material that could cause an adverse impact on the health and safety of workers or the public.

The intent of applying the defense-in-depth principles to each protective strategy is to ensure that defense-in-depth is considered in each line of defense (i.e., protective strategy), as well as in a broad sense across the entire design via the protective strategies.

Due to the importance of defense-in-depth, the results of applying the defense-in-depth principles to each of the protective strategies is summarized in Table 8-2. The basis for this table is discussed in Appendix G. Each of the items identified in Table 8-2 results in a topic for which a requirement is needed.

Table 8-2   Defense-in-depth provisions.

Defense-in-depth	Physical Protection	Stable Operation	Protective Systems	Barrier Integrity	Protective Actions
(1) Consider intentional and inadvertent events	Integral Design Process	Integral Design Process	Integral Design Process	Integral Design Process	Integral Design Process
(2) Consider prevention and mitigation in design	Security Assessment and Security Performance Standards	Cumulative limit on initiating event frequencies.	Accident prevention and mitigation • fuel damage criterion • coolable geometry criterion	Accident prevention and mitigation • barrier integrity criterion	Develop Emergency Operating Procedures and Accident Management integral with design EP

Table 8-2   Defense-in-depth provisions.

Defense-in-depth	Physical Protection	Stable Operation	Protective Systems	Barrier Integrity	Protective Actions
(3) Not dependent upon a single element of design, construction, maintenance, operation	Security Assessment and Security Performance Standards	Ensure events that can fail multiple protective strategies are $<10^{-7}$/plant year	Provide 2 independent, redundant diverse means for: reactor shutdown and decay heat removal	Provide at least 2 barriers	No key safety function dependent upon a single human action or piece of hardware
(4) Account for uncertainties in performance and provide safety margins	Security Assessment and Security Performance Standards	Reliability Assurance Program Provide safety margins in performance limits.	Reliability and availability goals consistent with assumptions in the PRA Reliability Assurance Program Use conservative source term Provide safety margin in regulatory limits	Provide radiological containment functional capability independent from fuel and RCS Use conservative source term Provide safety margin in regulatory limits	EP For safety margin, use conservative source term
(5) Prevent unacceptable release of radioactive material	Security Assessment and Security Performance Standards	Ensure events that can fail multiple protective strategies are $<10^{-7}$/plant year	N/A	Provide radiological containment functional capability independent from fuel and RCS	Accident Management
(6) Siting	Security Assessment and Security Performance Standards	Limits on external event cumulative frequencies	N/A	N/A	EP

N/A = Not applicable

8.2.5 Probabilistic Considerations (Figure 8-2: Box 5)

Chapter 6 identifies a number of probabilistic considerations that are key to ensuring a risk-informed licensing process consistent with the Commission's safety goals. This includes:

*   overall plant risk
*   a frequency-consequence (F-C) curve
*   probabilistic event selection criteria
*   criteria for the selection of LBEs

- LBE acceptance criteria
- risk-informed safety classification criteria
- risk-informed security performance standards

Box 5 of Figure 8-2 is intended to ensure that the answers to the questions identified in Box 3 incorporate the risk and design criteria contained in Chapter 6 of this document. Likewise, Chapter 7 identifies criteria for the scope and quality of the PRA necessary to support application of the risk-informed licensing process discussed in this NUREG. The answers to questions identified in Box 3 also need to incorporate these criteria.

8.2.6   Topics for Requirements (Figure 8-2: Box 6)

The answers to the questions for each protective strategy (Box 6 of Figure 8-2) leads to the identification of specific topics that the requirements will need to address to ensure adequate implementation of the protective strategies. These specific topics define the scope and content of the risk-informed technical and administrative requirements. Table 8-3 illustrates how the answers to the questions shown in Table 8-1 can be used to identify those topics. These topics are intended to ensure good practices and defense-in-depth are used in design and operation.

Once identified, the topics have been categorized as to whether they apply to design, construction or operation of the facility. Specifically,

- design refers to all engineering and analysis activities;

- construction refers to all on-site fabrication activities or off-site manufacturing activities that result in physical changes to the facility or material brought on-site for fabrication of the facility or for use over the life of the facility (e.g., fuel, spare parts, plant modifications, etc.)

- operation refers to all on-site operator and maintenance activities to startup, control and shutdown the facility beginning with initial fuel load and continuing through termination of power generation and preparation for decommissioning.

It should be noted that design, construction and operations activities can occur over the life of the plant and simultaneous with each other. As an example, Table 8-3 lists the topics resulting from the answers given to the questions in Table 8-1.

Table 8-3   Example topics resulting from answers to questions shown in Table 8-1.

Design Good Practices	Construction Good Practices	Operation Good Practices
Good Practices • Human Factors • Human Machine Interface • Verification of Analytical Tools • Quality of Safety Analysis and PRA **Defense-in-depth Considerations** • Provide at least 2 barriers to the release of FP • Independent radiological containment functional capability	**Good Practices** • QA / QC	**Good Practices** • Verify procedures • Trained personnel • Use of Simulator • Work control • Surveillance program • ISI program • IST program • Technical specifications • Monitoring and feedback **Defense-in-depth Considerations** • EP

The identification of the complete set of topics (i.e., implementation of Boxes 1 through 6) is described in Appendix G. Table 8-4 summarizes the results of that application, as documented in Appendix G. The topics contained in Table 8-4 result from applying the process shown in Figure 8-2 to all five of the protective strategies. The topics represent items for which requirements need to be written. These topics are organized by:

• topics common to design, construction, operation
• physical protection
• design
• construction
• operation
• administrative

Also shown in Table 8-4 is the location in the Framework where additional discussion related to each topic is provided. The discussion in Appendix G and the reference in Table 8-4 to the appropriate sections in Chapters 2 through 7, provide guidance as to the technical content that the requirements should contain, This, in conjunction with the guidance contained in Section 8.3, can be used to develop the requirements.

Table 8-4   Topics for which requirements are needed.

Topic	Framework Technical Description
General Topics Common to Design, Construction, and Operation	
• QA/QC	Appendix G - Section G.2.2
• PRA scope and technical acceptability	Chapter 7 and Appendix F
• Use of risk information	All chapters
• Integration of Safety, Security and Preparedness	Chapter 3
Physical Protection	
• General (10 CFR 73)	Appendix G - Section G.2.1
• Security performance standards	Section 6.7
Good Design Practices	
• Plant Risk: — Frequency-Consequence curve — Quantitative health objectives (including integrated risk)	Chapter 6
• Criteria for selection of LBEs	Chapter 6
• LBE acceptance criteria: — frequent events (dose, plant damage) — infrequent events (dose, plant damage) — rare events (dose) — link to siting	Chapter 6
• Initiating Event Severity (Keep initiating events with potential to defeat two or more protective strategies $<10^{-7}$/plant year)	Appendix G - Section G.2.2
• Criteria for safety classification and special treatment	Chapter 6
• Equipment Qualification	Appendix G - Section G.2.2
• Analysis guidelines — realistic analysis, including failure assumptions — source term	Chapter 6
• Siting and site-specific considerations	Appendix G - Section G.2.2
• Use consensus design codes and standards	Appendix G - Section G.2.2
• Materials qualification	Appendix G - Section G.2.2
• Protection against natural phenomena	Appendix G - Section G.2.2
• Dynamic effects	Appendix G - Section G.2.2
• Sharing of structures, systems, and components	Appendix G - Section G.2.2

Table 8-4   Topics for which requirements are needed.

Topic	Framework Technical Description
• Provide 2 redundant, diverse, independent means for reactor shutdown and decay heat removal	Appendix G - Section G.2.3
• Minimum - 2 barriers to FP release	Appendix G - Section G.2.3
• Radiological containment functional capability	Appendix G - Section G.2.4
• Radiological containment atmosphere cleanup	Appendix G - Section G.2.4
• Fracture prevention of radiological containment pressure boundary	Appendix G - Section G.2.4
• Electrical power systems	Appendix G - Section G.2.2
• Piping systems penetrating radiological containment boundary	Appendix G - Section G.2.4
• Closed system isolation valves	Appendix G - Section G.2.4
• Vulnerability to a single human action or hardware failure	Appendix G - Section G.2.5
• Need to consider degradation and aging mechanisms in design	Appendix G - Section G.2.2
• Reactor inherent protection (i.e., no positive power coefficient, limit control rod worth, stability, etc.)	Appendix G - Section G.2.2
• Human factors considerations	Appendix G - Section G.2.2
• Fire protection	Appendix G - Section G.2.2
• Control room design	Appendix G - Section G.2.5
• Alternate shutdown location	Appendix G - Section G.2.5
• Reactor core flow blockage and bypass prevention	Appendix G - Section G.2.2
• Specify reliability and availability goals consistent with PRA: — establish Reliability Assurance Program — specify goals on initiating event frequency	Appendix G - Section G.2.2
• Research and Development	Appendix G - Section G.2.2
• Use of prototype testing	Appendix G - Section G.2.2
• Combustible gas control	Appendix G - Section G.2.3
• Energetic reaction control	Appendix G - Section G.2.3
• Prevention of reactor coolant boundary brittle fracture	Appendix G - Section G.2.2
• Reactor coolant pressure boundary	Appendix G - Section G.2.2
• Reactor coolant activity monitoring and cleanup	Appendix G - Section G.2.3

Table 8-4   Topics for which requirements are needed.

Topic	Framework Technical Description

• I&C System
    — analog
    — digital
    — Human-machine interface
	Appendix G - Section G.2.2

• Protection of operating staff during accidents	Appendix G - Section G.2.5
• Control of release of radioactive materials to the environment	Appendix G - Section G.2.4
• Monitoring radioactivity releases	Appendix G - Section G.2.4
• Qualified analysis tools	Chapter 6
Good Construction Practices	
• Use accepted codes, standards, practices	Appendix G - Section G.2.2
• Security	Appendix G - Section G.2.1
• Non-destructive examination	Appendix G - Section G.2.2
• Inspection	Appendix G - Section G.2.2
• Testing	Appendix G - Section G.2.2
Good Operating Practices	
• Radiation protection during routine operation	Appendix G - Section G.2.2
• Maintenance program	Appendix G - Section G.2.2
• Personnel qualification	Appendix G - Section G.2.2
• Training	Appendix G - Section G.2.2
• Use of Procedures	Appendix G - Section G.2.2
• Use of simulators	Appendix G - Section G.2.2
• Staffing	Appendix G - Section G.2.2
• Aging management program	Appendix G - Section G.2.2
• Surveillance (including materials surveillance program)	Appendix G - Section G.2.2
• ISI	Appendix G - Section G.2.2
• Testing	Appendix G - Section G.2.2
• Technical specifications, including environmental	Appendix G - Section G.2.2
• EP	Appendix G - Section G.2.5
• Monitoring and feedback	Appendix G - Section G.2.2

Table 8-4   Topics for which requirements are needed.

Topic	Framework Technical Description
• Work and configuration control	Appendix G - Section G.2.2
• Maintenance of the PRA	Chapter 7
• Maintain fuel and replacement part quality	Appendix G - Section G.2.2
• Security	Appendix G - Section G.2.1
Administrative	
• Standard format and content of applications	Appendix G - Section G.3
• Change control process	Appendix G - Section G.3
• Record keeping	Appendix G - Section G.3
• Documentation control	Appendix G - Section G.3
• Reporting	Appendix G - Section G.3
• Corrective action program	Appendix G - Section G.3
• Backfitting	Appendix G - Section G.3
• License amendments	Appendix G - Section G.3
• Exemptions	Appendix G - Section G.3
• Other administrative items from 10 CFR 50	Appendix G - Section G.3

8.3   Development of Requirements

The next step of the process is the actual development of the requirements. This section provides guidance on how to take the topics identified in Section 8.2 (Box 6 on Figure 8-2) and develop requirements. The guidance covers those factors which need to be considered in developing the requirements. These factors are:

- use of 10 CFR 50 requirements, including general design criteria and their supporting regulator guides where practical (i.e., risk-informed and performance-based);

- use of a risk-informed and performance-based approach;

- ensure requirements consider lessons learned from the past; and

- perform a completeness check.

Section 8.2.6 identifies the topics which the requirements need to address to ensure the protective strategies are effectively implemented and appropriate administrative controls are in place. It is envisioned that the requirements be written to identify the broad principles and objectives associated with the requirements. As such, the general design criteria (GDC) currently contained in 10 CFR 50, Appendix A, serve as a good model of the scope, approach and depth envisioned in the requirements. In fact, it may be possible to use some of the existing GDCs directly as requirements. In some cases, more specificity may be needed in some requirements where specific criteria or design features are considered necessary. That is, in some cases, modifications to existing requirements may be appropriate; however, in other cases, new requirements may be necessary to address the diverse reactor technologies and to implement a risk-informed and performance-based approach. In any case, the factors listed above should be considered in writing the requirements so as to ensure completeness and consistency in their scope and approach.

Each of these factors is discussed in the sections that follow and their relationship to the requirements development process is shown in Figure 8-4. The development of the requirements will likely be an iterative process and should be guided by a group of experts serving in an advisory capacity.

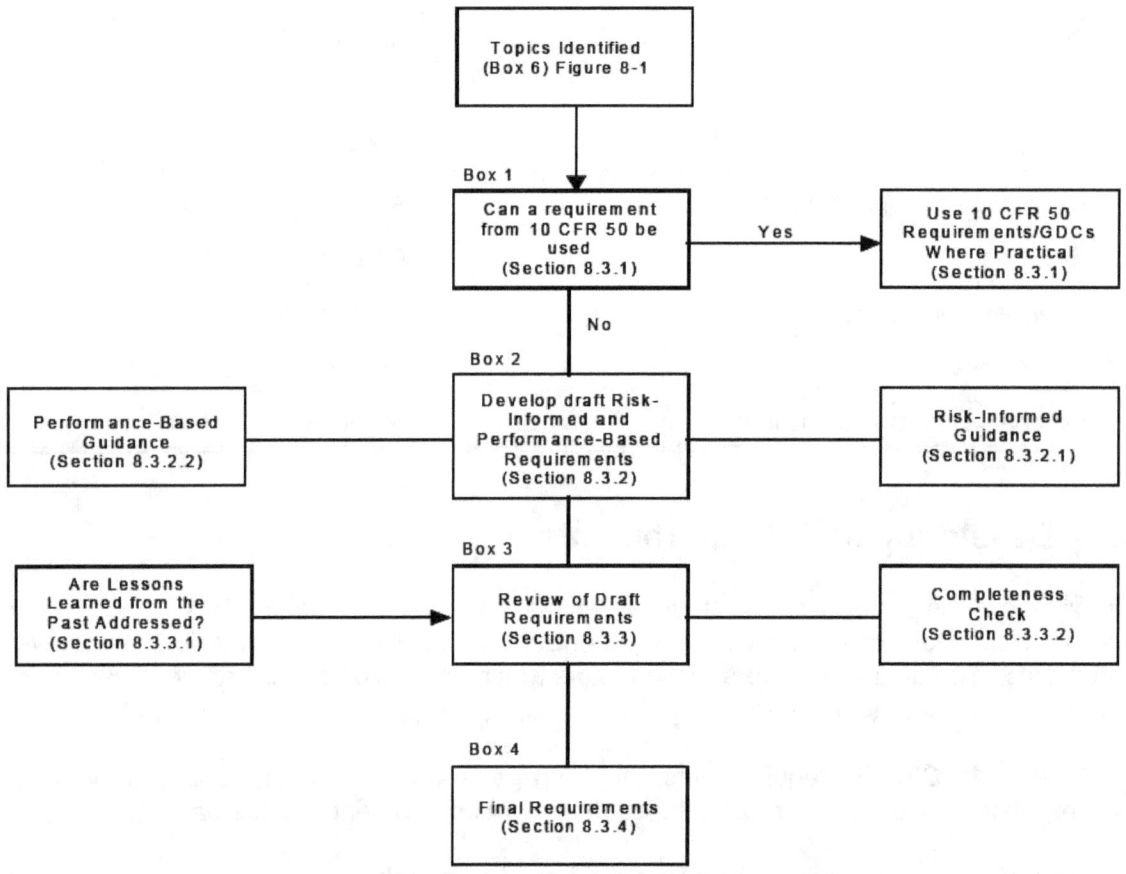

Figure 8-4   Development of Requirements

8.3.1   Use of 10 CFR 50 and Supporting Regulatory Guides (Figure 8-4:Box 1)

10 CFR 50 contains many requirements (both technical and administrative) and general design criteria that would be applicable. In developing the requirements for this NUREG, it would be most effective to use existing requirements and general design criteria whenever possible, since there

is experience in their use and any needed implementing guidance has already been developed. Accordingly, as a first step, 10 CFR 50 should be examined to see where its requirements and general design criteria could be used. In assessing whether or not to utilize existing wording from 10 CFR 50 in developing the requirements, the following approach should be followed:

- Is the wording acceptable as is (i.e., is it compatible with a risk-informed and performance-based approach)? If so, use the exact same wording as is in 10 CFR 50 or the general design criteria.

- Can the wording be used with some modification (i.e., can it be made more risk-informed or more performance-based)? If so, suggest the appropriate modifications.

- Is the wording not usable (i.e., too technology-specific or prescriptive)? If so, do not use. Develop a new requirement.

A preliminary examination of 10 CFR 50 has been conducted as described in Appendix H to identify which of the 10 CFR 50 requirements and general design criteria are candidates for use in a risk-informed and performance-based licensing approach. Where they can be used they should be used, along with their regulatory guides, providing the guides are also compatible with a risk-informed and performance-based licensing approach. Where they cannot be used, modifications may be appropriate, or new requirements needed.

8.3.2   Risk-Informed and Performance-Based Guidance (Figure 8-4: Box 2)

The development of the requirements is intended to result in a stand alone risk-informed and performance-based requirement that can serve as an alternative to 10 CFR 50 for licensing future plants. As such, the requirements should be based upon the guidance in Chapters 2 through 7 of this NUREG, including utilization of appropriate existing requirements and general design criteria as discussed above. Table 8-4 lists the technical and administrative topics identified from applying the process shown in Figure 8-2. The requirements need to address each of these topics in a risk-informed and performance-based fashion.

8.3.2.1   Risk-Informed Guidance

The development of the requirements is intended to result in a stand alone set of risk-informed and performance-based requirements that can serve as an alternative to 10 CFR 50 for licensing future plants. As such, the requirements should be based upon the guidance in Chapters 2 through 7 of this NUREG, including utilization of appropriate existing 10 CFR 50 requirements and general design criteria, as discussed in Section 8.3.1. Table 8-4 lists the technical and administrative topics identified as a result of applying the process shown in Figure 8-2. The requirements need to address each of these topics in a risk-informed and performance-based fashion.

A risk-informed and performance-based approach is one in which the risk insights, and engineering analysis and performance history are used to: (1) focus attention on the most important activities; (2) establish objective criteria based upon risk insights for evaluating performance; (3) develop measurable or calculable parameters for monitoring system and licensee performance; and (4) focus on the results as the primary basis of regulatory decision-making. Accordingly, guidance applicable to the above is discussed in the following sections.

In licensing using a risk-informed approach, a design specific PRA will play a central role in assessing the safety of the design, both in the initial licensing review and over the life of the plant. Accordingly, it is important that the requirements be written to be compatible, as much as possible, with the type of information that a PRA can provide. This means that in writing the requirements, they should be compatible with the risk measures, criteria and other information that PRAs produce, consistent with the guidance in this document. In addition, deterministic criteria should be used in selected areas (e.g., defense-in-depth) so as to make the requirements risk-informed, not risk based. Chapters 4, 6 and 7 contain guidance and criteria related to risk-informing the licensing approach, which the requirements need to address.

8.3.2.2 Performance-Based Guidance

A performance-based approach brings about a focus on results as the primary basis for regulatory decision making, whether PRA information is available or not.

A performance-based approach is characterized and recognized by the occurrence of five defined attributes. These attributes are:

(1)   A framework exists or can be developed to show that performance, as indicated by identified parameters, will serve to accomplish desired goals and objectives.

(2)   Measurable, calculable, or constructable parameters to monitor acceptable plant and licensee performance exist or can be developed.

(3)   Objective criteria to assess performance exist or can be developed.

(4)   Margins of performance exist such that if performance criteria are not met, an immediate safety concern will not result.

(5)   Licensee flexibility in meeting the established performance criteria exists or can be developed.

Appendix I provides additional guidance on the application of these attributes in developing performance-based requirements.

Another important aspect of performance-based requirements is the selection of parameters to be monitored. Performance-based requirements will require the monitoring of parameters that can be tied directly to the objectives of the requirements. Examples of performance-based parameters are:

- temperature
- pressure
- flow-rate
- fluid level
- radiation level
- voltage
- current

The success criteria from the PRA can provide insights as to the parameters that need to be monitored. The frequency of monitoring, the instrumentation to be used, its calibration, accuracy and operability all need to be considered. It is likely that any instrumentation necessary for monitoring performance-based requirements will be classified as safety significant and included in technical specifications.

8.3.3 Review of Potential Requirements (Figure 8-4: Box 3)

This NUREG provides guidance regarding factors to consider when developing the requirements. As the requirements are developed, they will need to be checked for conformance with this NUREG. Specific criteria should be developed to guide the check and should include the following questions:

- Are all the topics identified in the Framework addressed?
- Have the requirements addressed all of the criteria and guidance in the Framework?
- Have 10 CFR 50 requirements and general design criteria been used to the extent practical?
- Are the requirements risk-informed? (see Section 8.3.2.1)
- Are the requirements performance-based? (see Section 8.3.2.2)

Each of these criteria should be used to check the draft requirements after they are developed to ensure they are consistent with the intent of this NUREG. In addition, two other considerations need to be included in the review as discussed below.

8.3.3.1 Lessons Learned from the Past

Lessons learned from the past should be applied in writing the requirements and their implementing guidance. Gaps in the requirements or guidance should not be repeated where they have lead to problems or safety concerns. Examples of lessons learned that should be addressed include:

- importance of using good engineering, construction and operating practices

- importance of containment to account for uncertainties (i.e., defense-in-depth)

- containment design issues (e.g., preventing direct impingement of molten core material on the containment shell)

- preventing flow blockage

- no positive power coefficient

- limiting control rod worth

- plant aging phenomena and locations

- stress corrosion and other forms of cracking (e.g., fatigue)

- chemical and flow induced erosion

- flow induced vibration

- items identified by the NRC generic safety issues program as documented in NUREG-0933

- items identified by the Electric Power Research Institute (EPRI) in their Advanced Light-Water Reactor (ALWR) Utility Requirements Document.

A thorough check of the draft requirements with respect to the above is needed.

8.3.3.2 Completeness Check

A check has been made on the completeness of the topics listed in Table 8-4. This check consisted of comparing the topics in these tables to the content of 10 CFR 50, the International Atomic Energy Agency (IAEA) safety standards for nuclear reactor design [IAEA 2000a] and operation [IAEA 2000b] and Nuclear Energy Institute (NEI) document 02-02 [NEI 2002], and the U.K. Health and Safety Executive document on safety assessment principles [HSE 2006]. These documents represent stand-alone sets of requirements applicable to nuclear reactor design, construction and operation. The purpose of the review was to help ensure that this document has identified all technical topics necessary for safety and all administrative topics necessary for licensing and regulatory oversight. The results of the review are documented in Appendix K of this NUREG.

Based upon the completeness check, the following items were included in the IAEA and U.K. reports, but not identified in the Framework:

- automatic safety actions in initial stage of accidents (does not require operator actions in the initial plant response to an accident)

- escape routes (provides for design features to protect operating staff)

- design fuel assemblies to permit inspection (requires fuel assembly designs to allow visual inspection for fuel pin and assembly damage prior to and after being in the reactor)

Each of these items will be reviewed as part of the implementation of the Framework (as discussed in Chapter 9 and Appendix C) and assessed as to whether or not they should be incorporated.

8.3.4 Potential Requirements (Figure 8-4: Box 4)

Applying the guidelines above to the topics identified in Table 8-4 should result in a set of requirements that are risk-informed and performance-based alternatives to 10 CFR 50 for licensing new nuclear power plants. Appendix J contains a set of potential requirements developed using the above guidance.

One of these potential requirements, resulting from the examples in Table 8-3, is discussed below with respect to the four factors at the beginning of Section 8.3 to illustrate application of the guidance.

Potential Example Requirement

One potential example requirement addresses Qualified Analysis Tools, which is in Appendix J, Table J-5, as Design Requirement #42[8]:

> The analysis tools used in the licensing analysis shall be qualified for use by validation against data obtained from acceptable test programs and/or actual operating experience. The analytical tools shall be shown to be validated for use over the range of conditions expected and shall be capable of quantifying uncertainties. The analysis tools, test data, program description and their validation process and results shall be submitted to NRC for review.

Application of Factors

- Use 10 CFR 50 Requirements and General Design Criteria, Where Practical - this potential requirement is intended to ensure that the analytical tools used in the applicant's licensing analysis are valid for the conditions being analyzed. No equivalent requirement currently exists in 10 CFR 50 and, therefore, a new requirement is needed.

- Risk-Informed and Performance-Based - this potential requirement applies to the licensing analysis, including the PRA.

- Lessons Learned from the Past - this potential requirement addresses experience from the analysis of the AP-600 passive safety features where the applicant's and NRC's analysis tools needed to be qualified for the AP-600 conditions.

- Completeness - this potential requirement is included in principle in other design requirement documents.

Finally, once the requirements are developed, there remain many steps prior to their implementation, including the resolution of policy and additional technical issues and development of implementing guidance. These steps are discussed in Chapter 9.

8.4   References

[IAEA 2000a]   International Atomic Energy Agency. NS-R-1: IAEA Safety Standards Series "Safety of Nuclear Power Plants: Design." Vienna, Austria. 2000a.

[IAEA 2000b]   International Atomic Energy Agency. NS-R-2: IAEA Safety Standards Series "Safety of Nuclear Power Plants: Operation." Vienna, Austria. 2000b.

[NEI 2002]   Nuclear Energy Institute. NEI-02-02: "A Risk-Informed, Performance-Based Regulatory Framework for Power Reactors." Washington, D.C. May 2002.

[HSE 2006]   U.K. Health and Safety Executive. "Safety Assessment Principals for Nuclear Facilities" 2006 edition. Bootle, Merseyside, United Kingdom. 2006.

---

[8]Where the draft example requirements use the words "will" or "shall", they are for the purpose of illustration only.

9. CONCLUSIONS AND IMPLEMENTATION

9.1 Introduction

The purpose of this chapter is to provide (1) an assessment of how the Framework objectives have been met, (2) an evaluation of the feasibility of developing a risk-informed and performance-based approach for future plant licensing, and (3) the steps necessary to fully implement the Framework. The use of a fully risk-informed and performance-based approach for licensing has the potential to improve safety by allowing the designers and regulators to focus on the most safety significant structures, systems and components (SSCs) and human actions, while at the same time providing the basis for establishing performance measures for monitoring plant safety.

This approach can result in the development of requirements that could be implemented on a design specific, technology-specific or technology-neutral basis. Implementation could be via design specific licensing reviews or rule-making. Draft example requirements, developed using the Framework guidance, are provided in Appendix J to illustrate the application of this NUREG for developing risk-informed and performance-based requirements for future plant licensing. However, it needs to be emphasized that this document does not represent a staff position, but rather a concept and demonstration of the feasibility of developing a risk-informed and performance-based licensing approach. As discussed in this chapter, there remains much work to be done to finalize this approach before using it in the licensing process if, and when, a decision is made to do so.

This NUREG is written for NRC staff use if, and when, it is decided to apply a fully risk-informed and performance-based licensing approach. The technical basis contained in this document includes information on (1) the overall approach to developing risk-informed and performance-based requirements, (2) guidance on how to include the use of probabilistic risk information in the requirements, (3) guidance on how to include defense-in-depth in the requirements, (4) guidance on how to select and write the requirements and acceptance criteria, including the use of existing requirements where practical and (5) a set of example draft requirements to illustrate the outcome of applying the Framework. This NUREG is also written so that it can be applied on a technology-specific or technology-neutral basis and implemented generically or on a design specific basis. The discussion below summarizes how the objectives of this effort have been met, the conclusion with respect of feasibility and the major steps, necessary for implementation, if, and when, a decision is made to continue work to support the development of a risk-informed and performance-based licensing approach.

9.1.1 Objectives

In Chapter 1, several objectives were identified that the Framework, when developed, was intended to achieve. Those can be summarized as follows:

- Risk-Informed — ensure that risk information is integrated into the licensing process;

- Performance-Based — ensure that the requirements developed using the Framework process and implemented are based upon using plant performance;

- Defense-in-Depth — ensure defense-in-depth is maintained and uncertainties are appropriately accounted for; and

- Flexible — ensure that the technical basis in the Framework can be applied to reactors of diverse design.

Achieving these objectives should result in a licensing process that is more effective, efficient and stable than the current LWR oriented licensing process contained in 10 CFR 50.

The factors used in judging whether or not each of the above objectives have been met are:

- Is the Framework consistent with Commission policies?

- Is the Framework compatible with other parts of 10 CFR?

- Is the Framework compatible with a one-step and a two-step licensing process and does it contain provisions for exemptions, if necessary?

- Does the Framework incorporate past experience and lessons learned from the many years of reactor regulation?

Applying the above factors to each of the objectives results in the following:

- **Risk-Informed —**

  The use of risk information as an integral part of future licensing requirements is consistent with the Commission's 1995 Policy Statement on the Use of PRA [NRC 1995], and their March 11, 1999 white paper on risk-informed and performance-based regulation [NRC 1999]. This NUREG includes deterministic elements (e.g., defense-in-depth, good engineering practices) that ensure the Framework is not risk-based. The risk-informed aspects of the Framework ensure that the focus is on the most safety-significant SSCs on a design-specific basis, thus facilitating effectiveness. This is done in a fashion that is compatible with other parts of 10 CFR (e.g., use of existing dose criteria, siting criteria, etc.) and can be applied in a one-step or two-step licensing process. Finally, the risk-informed aspects of the Framework incorporate experience with the use of risk information in the regulatory process (e.g., standards development for PRA quality, risk informing selected 10 CFR 50 requirements) and, incorporating into the draft example requirements contained in Appendix C measures that address lessons learned (see Chapter 8). Thus, it is concluded that the objective has been met.

- **Performance-Based —**

  By its very nature, the use of risk information in licensing needs to be done on a design specific basis. Demonstrating that risk criteria are met requires analysis, monitoring and feedback of plant performance parameters. The process and criteria contained in the Framework provide for design specific implementation through the use of plant performance measures. This is consistent with the Commission's March 11, 1999 white paper on risk-informed and performance-based regulation. The performance-based approach used in this NUREG does not impact compatibility with other parts of 10 CFR and can be used in a one-step or two-step licensing process. It also reflects lessons learned, particularly in facilitating development of a reactor oversight program that can be tied directly to the requirements. Thus, it is concluded that this objective has been met.

- **Defense-in-Depth —**

  The Commission has had a philosophy of defense-in-depth since the early days of reactor regulation. The Framework includes this philosophy and provides the basis for a proposed policy statement on the purpose and elements of defense-in-depth. This policy statement would assist in providing consistency and direction to the NRC staff, applicants, licensees and other stakeholders on the application of defense-in-depth in the regulatory process. The defense-in-depth approach applied in the Framework is compatible with other parts of 10 CFR and with a one-step or two-step licensing process. The Framework's basis for a policy statement on defense-in-depth has taken into consideration past NRC practices and discussion on this topic as well as international (e.g., IAEA) and other stakeholders (e.g., Nuclear Energy Institute (NEI)) views. Thus, it is concluded that this objective has been met.

- **Flexible —**

  The Commission, in the past, has licensed non-LWR reactor designs. However, this required exemptions to requirements in 10 CFR 50 and the development of new requirements specific to the non-LWR design being licensed. Thus, the Commission has accommodated flexibility in the licensing process; however, the Framework has been written to provide that flexibility without the need for (or with minimum need) exemptions or the development of design specific requirements, thus enhancing stability and efficiency in the licensing process. In this regard, this document retains the legal and process requirements contained in 10 CFR 50 since there is experience and guidance associated with their use. This has been done in a fashion compatible with other parts of 10 CFR and with a one-step or two-step licensing process. The requirements development process has a specific element that includes consideration of experience and lessons learned, as described in Chapter 8. Thus, it is concluded that this objective has been met.

Based upon the above, the authors conclude that the Framework's objectives have been met and, if properly implemented, can result in a potentially more effective, efficient and stable licensing process than that currently used.

9.1.2   Feasibility

The Framework is intended to provide sufficient guidance and the technical basis for establishing the feasibility of developing and implementing a fully risk-informed and performance-based approach for future plant licensing. The authors conclude that the Framework does establish this feasibility. This conclusion is based upon the following:

- Top down risk criteria have been developed that can be implemented using design-specific risk information.

- The risk criteria have been integrated with deterministic criteria (e.g., defense-in-depth) to ensure a risk-informed, not a risk-based approach.

- Processes for using design specific information to establish key elements of the plant licensing basis (e.g., licensing-basis events (LBEs), safety classification) have been developed and tested on an existing light-water reactor (LWR) design (see Appendix E). This test has shown the processes to be feasible.

- A definition and principles for defense-in-depth have been developed and applied in a fashion compatible with the use of risk information. They have led to the identification of deterministic requirements to account for uncertainties.

- Guidance related to the scope and quality of the probabilistic risk assessment (PRA) needed to implement the Framework process have been developed.

- A process has been developed and applied that leads to a set of risk-informed and performance-based requirements (see Appendix J for draft set of example requirements).

- The draft example requirements are based, as much as possible on existing dose criteria, interface with other parts of 10 CFR and utilize existing requirements, as much as practical.

- The draft example requirements in Appendix J have been compared to other documents containing design and operational requirements (e.g., International Atomic Energy Agency (IAEA) design requirements) and no significant gaps have been found.

Thus, the authors conclude that the Framework has demonstrated the feasibility of developing a risk-informed and performance-based approach for future plant licensing that addresses design, construction and operation. Although feasibility has been demonstrated, to accomplish full implementation, a number of follow-on actions will be needed. These follow-on actions include:

- programmatic direction for implementation;

- resolution of policy and open technical issues;

- completion of implementation steps:

  — development of draft requirements;

  — development of implementing guidance for new requirements and identification of applicable existing guidance;

  — pilot testing the draft requirements and implementing guidance on a non-LWR design;

  — development of a compatible reactor oversight program; and

  — rule-making.

Each of these actions is discussed briefly below so as to provide a roadmap to the elements of implementation.

9.2   Programmatic Direction for Implementation

The method by which the Framework is to be implemented will be a key factor in setting the direction and in estimating the resources and schedule for additional work. Specifically, there are two fundamental issues that need to be resolved prior to implementation of the Framework. These are:

(a)   should the Framework be implemented by rule-making or on a case-by-case basis? and

(b)   should the Framework be implemented in a technology-specific or technology-neutral fashion?

Each of these issues have options for their resolution as discussed below:

(a)   **Rule-making versus Case-by-Case Implementation —**

The Framework has been written such that it can be implemented either through rule-making or on a design-specific case-by-case basis, without rule-making. In design-specific implementation, the requirements would be documented in the staff's Safety Evaluation Report or Final Design Approval and Design Certification. If rule-making is chosen as the path for implementation, it could be technology-specific or technology-neutral and could be accomplished via modification of, or supplementing 10 CFR 50, or adding a stand alone new part to 10 CFR. The Framework has been written to support a stand alone implementation but could also support implementation via modification to 10 CFR 50. In SECY-07-0101 [NRC 2007a] the staff recommended deferring a decision on rule-making until after development of the licensing strategy for Next Generation Nuclear Plants (NGNP) or receipt of an application for design certification of the Pebble Bed Modular Reactor (PBMR). The Commission, in a September 10, 2007, Staff Requirements Memorandum (SRM) [NRC 2007b], agreed with the staff recommendation and suggested that the Framework be tested on a non-LWR, such as the PBMR, prior to a decision on rule-making. Such a test can be done with technology-specific requirements and guidance or with technology-neutral requirements and technology-specific implementation. Accordingly, it may be several years before the issue of rule-making is decided, including whether it should be technology-specific or technology-neutral as discussed below.

(b)   **Technology-Specific versus Technology-Neutral Implementation —**

Resolution of the policy and open technical issues described in Section 9.3 is dependent upon whether or not they are being viewed from a technology-specific or technology-neutral standpoint. The Framework has been written such that it can be implemented in either a technology-specific or technology-neutral fashion.   Specifically, the draft example requirements in Appendix J of this NUREG have been written in a technology-neutral fashion and those areas where technology-specific guidance will be necessary are identified. In addition, Appendix J illustrates the content of the technology-specific guidance that will be needed. However, the resolution of policy and open technical issues may be more straight forward if done in a technology-specific fashion since only the safety issues, technical basis and uncertainties associated with that technology will need to be considered. Also, only one set of implementing guidance will need to be developed, in lieu of multiple sets to cover other technologies. This also will likely be more straight forward and quicker than a technology-neutral approach.

This NUREG takes no position on the above issues, but rather identifies them as over-arching issues needing resolution prior to any significant additional work on implementation.

9.3   Resolution of Policy and Open Technical Issues

In developing the Framework, a number of issues of a policy nature (i.e., major changes in current Commission regulatory requirements, approaches or practices) have been identified. Seven of these issues were originally identified in SECY-03-0047 [NRC 2003] and one in SECY-05-0120 [NRC 2005] (Security Performance Standards), as indicated below, and are discussed in Appendix C. Listed below are the policy issues associated with the Framework:

- Defense-in-Depth
- Level of Safety
- Integrated Risk
- Probabilistic Licensing Basis
- Source Term
- Containment
- Emergency Planning
- Security Performance Standards

Although the Commission had previously approved the staff's recommendation to pursue resolution of these issues, the details of their resolution will require Commission review and direction prior to finalizing the risk-informed and performance-based requirements and their implementing guidance. Accordingly, this NUREG has been written to reflect a recommended resolution of each policy issue as described in Appendix C. If the Commission directs a resolution to one or more of the policy issues different than that taken in the Framework, then this NUREG will need to be reevaluated and modified accordingly.

Similar to the policy issues, there are a number of open technical issues that need to be resolved as part of Framework implementation. The major open technical issues are also described in Appendix C and consist of:

- Complementary Cumulative Distribution Function (CCDF) — Is a cumulative risk curve needed?

- Frequency-Consequence Curve — Are the frequencies, dose and structure used satisfactory?

- Fuel Handling and Storage — What criteria and requirements should apply?

- Environmental Protection — Do the Framework criteria for safety also adequately protect the environment?

- Framework Testing — How should testing against an actual design be accomplished?

- Security Frequency-Consequence Curve — Should the consequences be specified as dose or individual risk?

- Design Codes and Standards — What consensus design codes and standards are needed for non-LWRs and when should they be developed?

- PRA Standards and Use of PRA — What standards are needed for non-LWR PRA quality and what guidance is needed for use of the PRA in licensing?

- Subsidiary Risk Objectives — Are they needed and how should they be developed for non-LWRs?

- Importance Measures — What measures are needed, how should they be derived used?

- Completeness Check Findings — How should the findings be addressed?

- Reactor Oversight Process (ROP) — What ROP elements are needed for consistency with a risk-informed and performance-based licensing approach?

9.4   Implementation Steps

For full implementation of a risk-informed and performance-based licensing approach, based upon the Framework, the following steps will likely be necessary.

9.4.1   Development of Draft Requirements

Appendix J contains a set of example draft requirements, developed to illustrate what a set of risk-informed and performance-based requirements could look like following the Framework guidance. These could be used as the starting point for further development of requirements to support implementation. It is recognized that the example requirements contained in Appendix J could be affected by the resolution of the policy and open technical issues described above.

9.4.2   Development of Implementing Guidance

Some of the requirements will need technology-specific implementing guidance. For others, technology-neutral guidance will be sufficient. For example, a requirement to maintain "coolable geometry" will require technology-specific guidance as to what represents a loss of coolable geometry. Accordingly, it is envisioned that both technology-specific and technology-neutral regulatory guidance will be necessary to implement the requirements.

Appendix J identifies, for each example draft requirement, the need for implementing guidance. As identified in Appendix J, this guidance would be technology-neutral in nature but, in some cases, would also need to be technology-specific. Brief statements are also given in Appendix J regarding the scope and, in some cases, a source that could be used for the guidance. Appendix J could be used as the starting point for the development of guidance needed to implement the requirements.

9.4.3   Pilot Testing

A number of stakeholders suggested that before formally using the Framework in any licensing or rule-making actions, it should be pilot tested using an actual reactor design. The authors of the Framework consider this a critical step which should be done to confirm the practicality, completeness, clarity and acceptability of the requirements and their implementing guidance. Appendix E describes a limited test of the LBE selection process using an existing pressurized water reactor (PWR) PRA. This Appendix demonstrated that the LBE selection process can result in a more realistic set of design-basis accidents (DBAs). However, a more comprehensive pilot test is needed to examine all aspects of the Framework. The results would then be used to modify the requirements and their implementing guidance prior to use in any formal licensing action. This is discussed further in Appendix C as an open technical issue.

It should be noted that the Commission, in an SRM dated September 10, 2007, indicated that the Framework should be tested on an actual design and suggested the PBMR as a logical choice.

9.4.4   Reactor Oversight Process (ROP)

Over the life of a nuclear power plant, NRC has an oversight responsibility to ensure the plant is operated and maintained in accordance with its license.   For a fully risk-informed and performance-based licensing approach, an ROP will need to be developed and implemented that is consistent with the approach, metrics and performance measures used in the safety analysis. A fully risk-informed and performance-based licensing approach will facilitate the development of the ROP, since the performance measures can be derived directly from the risk analysis and its success criteria, which would be an integral part of the safety analysis. This is discussed further in Appendix C.

9.4.5   Rule-making

In addition to the above follow-on actions, if rule-making is chosen as the path for implementing the requirements, then additional follow-on actions would also be required. These include:

- development of regulatory analysis and an environmental assessment;

- development of a draft rule and supporting Regulatory Guides (RGs)/Standard Review Plans (SRPs) (The rule could be technology-specific or technology-neutral and this would likely be a policy decision); and

- identification of any conforming changes needed in other parts of 10 CFR, as well as other agency (e.g., Federal Emergency Management Agency (FEMA)) state and local regulations, policies and practices. This step would also be needed for design specific application of the risk-informed and performance-based requirements.

The above implementation steps will take time and resources and involve some iteration, but are considered necessary prior to using the Framework as the basis for any formal licensing action.

9.5   Summary

The authors conclude that the Framework has established the feasibility of developing a risk-informed and performance-based approach for future plant licensing. It has met the objectives laid out at the outset. As discussed in its September 10, 2007, SRM, the Commission has directed that the Framework be published and then tested on an actual design. The experience gained from such a test will provide valuable information for use in any further work on Framework implementation. Further work would include decisions on programmatic direction and resolution of the policy and open technical issues. The open nature of these issues, however, does not affect the validity of the information contained in this document.   Finally, the steps discussed in Section 9.4 are also considered necessary to fully implement a risk-informed, performance-based licensing approach.

Given the above, it is also important to emphasize that this NUREG, in its present form, could also be used by the NRC staff as a reference document to support:

- review of an applicant's technical basis for any proposal they may make to use risk information in the development of their licensing basis;

- developing questions for the applicant as part of reviewing their proposed risk-informed licensing approach; and

- development of a procedure for independently assessing the safety of a reactor design.

Therefore, in addition to its future value, this NUREG (in its present form) is considered a useful document that can support current or future staff activities.

9.6   References

[NRC 1995]   U.S. Nuclear Regulatory Commission, "Final Policy Statement 'Use of Probabilistic Risk Assessment (PRA) Methods in Nuclear Regulatory Activities'," Washington, DC, 60 FR 42622, August 1995.

[NRC 1999]   U.S. Nuclear Regulatory Commission, Yellow Announcement #019, "Commission Issuance of White Paper on Risk-Informed and Performance-Based Regulation," Shirley Ann Jackson, March 11, 1999.

[NRC 2003]   U.S. Nuclear Regulatory Commission, SECY-03-0047, "Policy Issues Related to Licensing Non-light-water Reactor Designs," March 28, 2003.

[NRC 2005]   U.S. Nuclear Regulatory Commission, SECY-05-0120, "Security Design Expectations for New Reactor Licensing Activities," July 6, 2005.

[NRC 2007a]   U.S. Nuclear Regulatory Commission, SECY-07-0101, "Staff Recommendations Regarding a Risk-Informed and Performance-Based Revision to 10 CFR Part 50," June 14, 2007.

[NRC 2007b]   U.S. Nuclear Regulatory Commission, "Staff Requirements - SECY-07-0101 - Staff Recommendations Regarding a Risk-informed and Performance-based Revision to 10 CFR Part 50," September 10, 2007.

www.ingramcontent.com/pod-product-compliance
Lightning Source LLC
Chambersburg PA
CBHW081447170526
45166CB00008B/2347